Flavonoid Metabolism - Recent Advances and Applications in Crop Breeding

Edited by Hafiz Muhammad Khalid Abbas and Aqeel Ahmad

Published in London, United Kingdom

Flavonoid Metabolism - Recent Advances and Applications in Crop Breeding
http://dx.doi.org/10.5772/intechopen.100875
Edited by Hafiz Muhammad Khalid Abbas and Aqeel Ahmad

Contributors
Soumya Basu, Shuchi Nagar, Saurabh Dey, Aishik Das, Ngoc Van Thi Nguyen, Shyamal K. K Jash, Indireddy Theja, Banoth Ramya Kuber, Asmaa Nabil-Adam, Mohamed E. Elnosary, Mohamed L. Ashour, Nehad M. Abd El-Moneam, Mohamed A. Shreadah, Shuchi Dave Mehta, Sukirti Upadhyay, Gopal Rai, Zorita Maria Diaconeasa, Andreea Stănilă, Daria Teodora Pop

First published in London, United Kingdom, 2023 by IntechOpen
IntechOpen is the global imprint of INTECHOPEN LIMITED, registered in England and Wales, registration number: 11086078, 5 Princes Gate Court, London, SW7 2QJ, United Kingdom

British Library Cataloguing-in-Publication Data
A catalogue record for this book is available from the British Library

Additional hard and PDF copies can be obtained from orders@intechopen.com

Flavonoid Metabolism - Recent Advances and Applications in Crop Breeding
Edited by Hafiz Muhammad Khalid Abbas and Aqeel Ahmad
p. cm.
Print ISBN 978-1-80356-704-4
Online ISBN 978-1-80356-705-1
eBook (PDF) ISBN 978-1-80356-706-8

Meet the editors

Hafiz Muhammad Khalid Abbas is a postdoctoral research associate working in the field of plant molecular biology at the University of Tennessee, USA. He earned his Ph.D. from Huazhong Agricultural University, Wuhan, China, and then joined the Guangdong Academy of Agricultural Sciences as a postdoctoral Fellow. He has undertaken an extensive study of plant defense responses and plant nutrient quality and uses functional genomics and metabolomics to understand plant-pathogen interactions. He has won one research grant and has published 19 research articles in leading journals including the *Journal of Agricultural and Food Chemistry, Theoretical and Applied Genetics, BMC Genomics,* and *Science China Life Sciences*. His editorial and review activities for multiple impact factor journals have made him a distinctive and progressive figure among his peers.

Dr. Aqeel Ahmad is research leader for crop cultivation under extreme climates at the Institute of Geographical Sciences and Natural Resources Research, China Academy of Sciences. He is an active researcher in plant stress physiology and hybrid coding for metadata analysis in crop modeling, and holds three registered patents.

Contents

Preface

Flavonoids are the largest and most diverse group of secondary metabolites found in plants, including fruits, vegetables, grains, and herbs. They play a variety of significant roles in determining plant growth and development, including UV protection, pollinator attraction, and defense against environmental stresses, specifically pests and diseases. Recently, scientists have conducted a significant amount of research to develop an understanding of the biosynthesis and regulation mechanisms of flavonoids, as well as their potential future applications in crop breeding.

The discovery of novel enzymes and genes involved in flavonoid biosynthesis pathways is a recent and important advance in the understanding of flavonoid metabolism. This has enabled scientists to manipulate these pathways in plants, improving overall plant health, including stress tolerance and disease resistance, and producing flavonoids with desirable properties such as improved antioxidant activity or increased pigment content to enhance the nutritional value of the specific crop. For example, researchers have been able to increase the flavonoid content of crops such as tomatoes, grapes, rice, and soybeans by overexpressing specific flavonoid biosynthesis genes.

Another area of research that has focused on elaborating on the role of flavonoids is plant responses to abiotic and biotic stresses. Flavonoids are well known to play central roles in protecting plants against environmental stresses such as UV radiation, drought, and a variety of pathogens. Through understanding how flavonoids contribute to stress tolerance, researchers are seeking to develop new strategies to improve crop yields and resilience in the face of changing environmental conditions.

Recently, growing interest has also been observed in studying the potential health benefits of flavonoids for humans. Flavonoids have been shown to possess a range of beneficial effects on human health, including antioxidant and anti-inflammatory activities, and they may also exhibit anticancer and cardiovascular protective properties. By increasing the flavonoid content of crops, it may be possible to improve and enhance the nutritional quality of food to provide additional health benefits to consumers.

Overall, recent advances in flavonoid metabolism research have significant implications for crop breeding and food production, playing a critical role in the development of healthier and more sustainable food crops, as well as new applications in the pharmaceutical and cosmetics industries.

The book consists of seven chapters.

Chapter 1 'Chemistry and Role of Flavonoids in Agriculture: A Recent Update', explains the overall importance of flavonoids in plant systems. It describes the detailed biosynthesis and chemistry of these metabolites inside the plants and the pathways they adopt to combat biotic and abiotic stress. The importance of flavonoids in determining the future of plant breeding with respect to food quality is highlighted.

Chapter 2, 'Importance of Flavonoid as Secondary Metabolites', discusses the anti-inflammatory, antioxidant, anticancer, anti-cholinesterase, antimicrobial, hepatoprotective, neuroprotective, cardioprotective, and antiallergic properties of flavonoids. Their scavenging activity enables superoxide, hydroxyl, and lipid radicals to play a crucial role against life-limiting diseases such as diabetes, cardiovascular disorder, and cancer.

Chapter 3, 'Flavonoids Biosynthesis in Plants as a Defense Mechanism: Role and Function Concerning Pharmacodynamics and Pharmacokinetic Properties', examines the biosynthesis pathway of flavonoids in plants and their role against abiotic and biotic stress factors. It also highlights the pharmacological functions of flavonoids and their impact on certain diseases, including dementia and Alzheimer's.

Chapter 4, 'Purple Corn Cob: Rich Source of Anthocyanins with Potential Application in the Food Industry', illustrates the value added by anthocyanins in the food coloring and beverage industry due to their antioxidant, anti-inflammatory and cardiovascular health benefits. Purple corn cob, a byproduct of corn, is cited as a specific example.

Chapter 5, 'Application of Liquid Chromatography in the Analysis of Flavonoid Metabolism in Plant', discusses the structural variability of different flavonoids and the use of methylation, glycosylation, acylation, and a further selection of different chromatographic techniques for analysis of specific compounds.

Chapter 6, 'Recent Advances in Flavonoid Metabolism: An Updated Review', describes the web of metabolic processes in which flavonoids are involved. It explains the influence of physical entities on flavonoid production and the regulatory mechanisms through which flavonoids function in plant cells. It also discusses minor entities and the interplay of various factors that are often ignored in flavonoid metabolism.

Chapter 7, 'Flavonoids: Recent Advances and Applications in Crop Breeding', describes the synchronized use of modern tools in breeding programs. The chapter explains the precise control exerted by flavonoids on plant reproduction and how it can be used to improve modern breeding techniques.

Hafiz Muhammad Khalid Abbas
Department of Plant Sciences,
University of Tennessee,
Knoxville, USA

Aqeel Ahmad
Institute of Geographic Sciences and Natural Resources Research,
Chinese Academy of Sciences,
Beijing, China

Chapter 1

Chemistry and Role of Flavonoids in Agriculture: A Recent Update

Shyamal K. Jash

Abstract

Flavonoids are a remarkable group of plant secondary metabolites, and are of importance and interest to a wide variety of physical and biological scientists. Continuing works on their chemistry, occurrence, natural distribution and biological function have already resulted a lot and have created a stir in the field of chemical and biological sciences due to their immense biological and pharmacological/therapeutic potential. Also flavonoids play an important role in the biological activities of plant system. They can be responsible for the color of flowers and fruits and for the attraction of pollinators. The plant flavonoids are used naturally to improve their adaptation to environmental stress, to improve food quality, and to increase crop yield. The present book chapter deals with chemistry and significance role of reported novel natural flavonoids along with a variety of activities in agriculture.

Keywords: naturally occurring flavonoids, biosynthesis, metabolism, chemistry of flavonoids, role of flavonoids in plants, agriculture, pest control, patent information

1. Introduction

Nature is an extremely rich source of highly diverse and innovative chemical structures with a variety of structural arrangement and interesting biological activities, which have played a significant role in the process of drug discovery and design. Chemistry of Natural Products has lately undergone explosive growth; natural products are of much interest and of promise in the present day research directed particularly toward drug-design and drug-discovery. Much research works were already carried out and also intensive works are now going on world-wide in the perspective of academic as well as pharmacological/therapeutic scenario. Statistically, only less than ~10–15% of the plants have been investigated so far; a major portion of them is still being left. To gather more knowledge on the natural availability of chemical compounds, their structural variety, properties, and the isolated compounds for detailed studies in regards to biological and pharmacological potentials, more and more research is demanded for the exploration of chemical nature of plants, particularly those ones which have been used as traditional medicines all over the world. Hence, the present investigator has been motivated to undertake this work on some plants traditionally used as medicine in India.

Research into secondary metabolism has long centered on flavonoids. Scientists from a wide variety of fields are interested and intrigued by flavonoids, which are widely found throughout the plant kingdom. Over the past few years, it has been revealed that plant flavonoids play a vital role in our lives and in the health of our plants. As a result of ongoing research on chemistry, occurrence, natural distribution and biological function of flavonoids, a number of reviews have already been published time to time [1–10]. A PubMed search incorporating the term "flavonoid" returns more than 58,180 articles from last 5 years. The role of flavonoids in plants has received considerable attention in recent decades [11].

Human culture is facing the greatest threat because of global climate change. Increasing global food prices and global warming put the future of humanity at risk. Scientists from NASA's Goddard Institute for Space Studies (GISS) estimate that global temperatures have risen by around 1°C since 1880 [12]. Every 2°C increase in global temperature could annihilate up to a hundred million people and wipe out up to a million species [13]. In addition to using fossil fuels to generate energy, agricultural activities are among the biggest contributors to climate change through the emission of greenhouse gases [14]. In spite of the convenience, ease of use, and rapid soil nutrient recharge, commercial fertilizers have become viewed as a source of toxic and residual soil issues. Using less mineral fertilizer may lower GHG emissions by 20% [15]. Global warming has made it necessary to rethink outdated and ineffective policies. Eco-friendly farming practices and a more sustainable agricultural system are urgently needed. Bio-based products, for example, might usher in organic farming, bio-fertilizers, and bio-control, all of which would be significant steps toward assuring global food security in the long run. Flavonoids are one type of biostimulant discussed in this chapter, and their role in sustainable agriculture. The flavonoids are an important class of polyphenolic secondary metabolites involved in plant physiological function, and show protection against biotic and abiotic stresses, including ultraviolet radiation, salt stress, and drought [16–18], at least in part by detoxifying the reactive oxygen species (ROS) produced when plants are under stress conditions [19]. Flavonoids may help protect Mediterranean endemic species from UV radiation and drought stress, as evidenced by recent studies that show polyphenol concentrations fluctuate monthly with the maximum values occurring at midday during the summer when drought, temperature, and UV radiation are high [20, 21]. The flavonoids in some plants play a critical role in plant defense and growth. There are several flavonoids that comprise plant pigments, including anthocyanins (red, orange, blue, and purple pigments); chalcones and aurones (yellow pigments); and flavonols and flavones (white and pale yellow pigments), which contribute to a diverse range of plant colors [22]. The flavonoids are also crucial in symbiotic associations between plants and microbes, such as rhizobial and arbuscular mycorrhizal symbioses [23]. As a signaling compound, certain flavonoids trigger the induction of nodule induction in rhizobia, which is the first step in legume-rhizobia symbiotic relationships [24]. In addition to preventing pests and pathogens, some flavonoids have antimicrobial properties [25]. The color pigments contained in some classes of flavonoids make leaf and flower petals distinctive, aiding plants in attracting pollinators [26]. Further, flavonoids have indirect effects on nutrient availability and supply since they enhance mycorrhizal symbiosis and enhance rhizosphere colonization by beneficial microbes [27].

2. Flavonoids: classification and biosynthesis network in plant

Flavonoids are the most diverse group of natural products; they are found in plants in over 10,000 different compounds [28]. In contrast to stilbenes (a class of

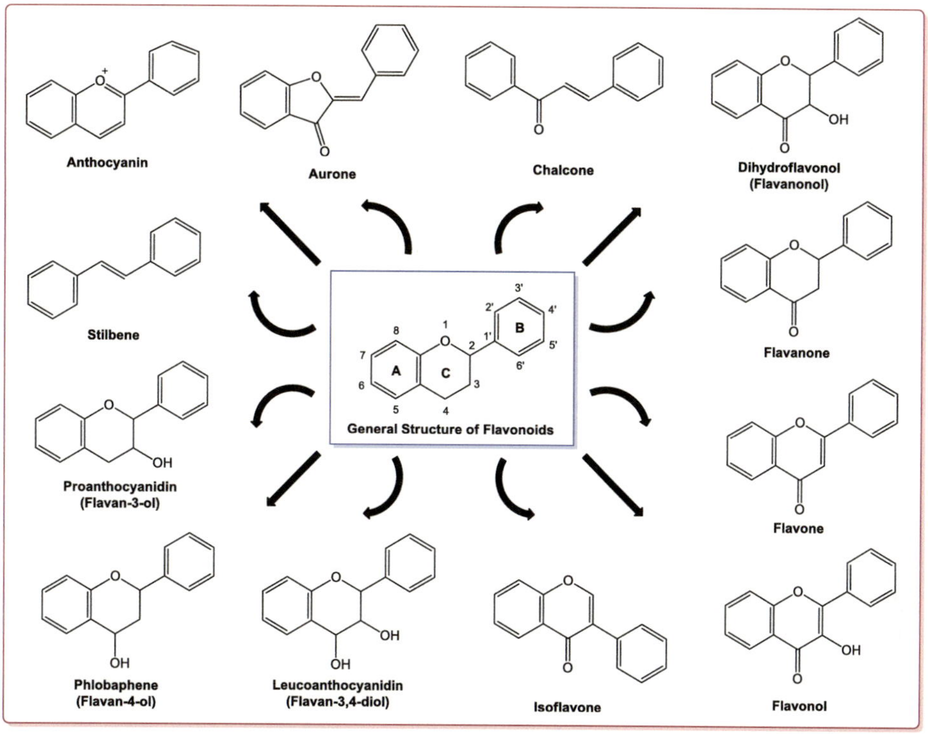

Figure 1.
Basic structure of flavonoids subclasses.

flavonoids) which has a C6-C2-C6 structure (**Figure 1**), flavonoids have a C6-C3-C6 basic structure composed of three phenolic rings, A (6 carbons) and B (6 carbons), linked by a 3-carbon heterocyclic ring (ring C). This structure, in turn, can give rise to several derivatives and sub-classes of compounds with distinct substituents [11, 29]. According to the degree of oxidation of the heterocyclic ring and the number of hydroxyl or methyl groups on the benzene ring, flavonoids can be divided into 12 subgroups: anthocyanins, aurones, chalcones, dihydroflavonols, flavanones, flavones, flavanols, isoflavones, leucoanthocyanidins, phlobaphenes, proanthocyanidins and stilbenes (**Figure 1**) [9, 30, 31]. However, in terms of attachment of the B ring to the C ring, flavonoids are into three main groups: Flavonoids (2-phenylbenzopyrans): the B ring is attached at the 2-position of the C ring, Isoflavonoids (3-phenylbenzopyrans): the B ring is attached at the 3-position of the C ring, and Neoflavonoids (4-phenyl-benzopyrans): the B ring is attached at position 4 of the ring C [11, 32].

The phenylpropanoid pathway produces flavonoids from phenylalanine, whereas the shikimate pathway produces phenylalanine [33]. It is generally recognized that the first three steps of the phenylpropanoid pathway are known as the general phenylpropanoid pathway [28]. Using this pathway, aromatic amino acid phenylalanine is transformed to top-coumaroyl-CoA via phenylalanine ammonia lyase (PAL), cinnamic acid 4-hydroxylase (C4H), and 4-coumarate: CoA ligase (4CL). A primary catalytic function of PAL is to catalyze deamination of phenylalanine to trans-cinnamic acid, the first in a general phenylpropanoid pathway [34]. Further, PAL is essential for the regulation of carbon flux from primary to secondary metabolism in plants [35]. StlA, which encodes PAL in *Photorhabdus luminescens*, has been shown to play a role in generating a stilbene antibiotic [34]. PAL activity has also been linked to anthocyanins and other phenolic compounds in strawberry fruit [36]. In the general phenylpropanoid pathway, the second step involves C4H, a monooxygenase found in cytochrome P450 and responsible for hydroxylating trans-cinnamic acid to generate p-coumaric acid. The flavonoid synthesis pathway involves this first oxidation reaction as well [37]. It has been found that the expression of C4H in *Populus trichocarpa* and *Arabidopsis thaliana* can be correlated with lignin content, an important phenylpropanoid metabolite [28]. The enzyme 4CL catalyzes the synthesis of p-coumaroyl-CoA by coupling with a co-enzyme A (CoA) unit to *p*-coumaric acid at the third step of the general phenylpropanoid pathway. A chalcone synthesizing enzyme, chalcone synthase (CHS), contributes to the biosynthesis of specific flavonoid-based compounds by combining one molecule of 4-coumaroyl-CoA (6-carbon) with three molecules of malonyl-CoA. As a result of two different pathways of cell metabolism, ring A and ring B are generated via the acetate pathway and shikimate

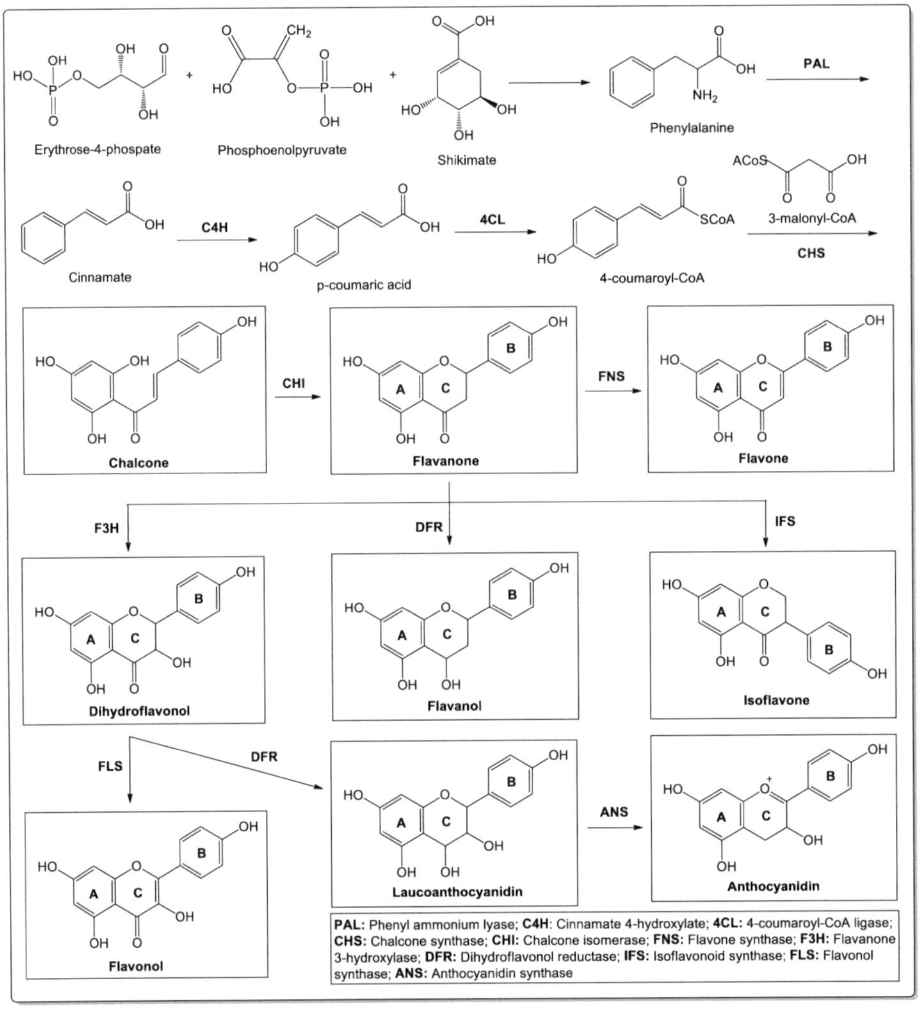

Figure 2.
Biosynthesis network of flavonoids in plant.

pathway, respectively, with chain linkages delivering ring C. During the acetate pathway, malonyl-CoA is converted to ring A by carboxylation of acetyl-CoA, whereas ring B and the linking chain (ring C) are generated via the shikimate pathway (**Figure 2**) from coumaroyl-CoA. In the phenylpropanoid pathway, coumaryl-CoA is directly generated by three enzymatic reactions from phenylalanine [29]. Following the condensation of these aromatic rings, these pathways lead to the synthesis of chalcone, which will then undergo isomerase-catalyzed cyclization to form flavanone (**Figure 2**). In addition to hydroxylation, glycosylation, and methylation, the latter compounds undergo additional modifications, resulting in an enormous variety of colors (**Figure 2**).

3. The production of flavonoids by microorganisms

Since plants and chemical synthesis produce low levels of flavonoids, researchers have turned to open fermentation and metabolic engineering to produce flavonoids in microorganisms [38]. Toxic chemicals and extreme reaction conditions are necessary for the chemical synthesis of flavonoids [39]. Combinatorial biosynthesis offers an advantage in the production of rare and expensive natural products, thanks to the rapid development of molecular biology tools and genome information flooding from a wide variety of organisms. Unlike the tedious blocking and de-blocking steps common to organic synthesis, it also allows for simple and complex transformations [40]. In addition to *Escherichia coli*, *Phellinus igniarius, Saccharomyces cerevisiae* and *Streptomyces venezuelae*, and, a medicinal mushroom, flavonoids can also be produced by additional prokaryotes and eukaryotes [41], a variety of other cultures have been used to produce flavonoids.

3.1 Phenylpropanoid pathway

Several flavonoids are synthesized in plants using the phenylpropanoid pathway from naringenin chalcone. A recently established biosynthesis pathway was established in a heterologous microorganism by fermentation of *E. coli* carrying an artificially assembled phenylpropanoid pathway to produce flavanones from amino acids such as phenylalanine and tyrosine [42]. Plants use phenylalanine ammonia lyase (PAL) to deaminate phenylalanine to produce cinnamic acid as the first step in the phenylpropanoid pathway. As a result of the action of cinnamate-4-hydroxylase (C4H), cinnamate-4-hydroxylase (C4H) converts cinnamate to p-coumaric acid, which is then converted to *p*-coumaroyl-CoA by 4-coumarate: CoA ligase, cinnamic acid becomes p-coumaric acid. The naringenin chalcone is synthesized by three acetate units from malonyl-CoA with p-coumaroyl-CoA using Chalcone Synthesis (CHS). In vitro, naringenin is converted to naringenin using chalcone isomerase (CHI) or nonenzymatically without activating enzymes [43].

3.2 Enhancement of flavonoid synthesis

For heterologous flavonoids production, many molecular biology technologies are used, including choosing promoter and target genes, knocking out related genes, over expressing malonyl-CoA, and creating artificial P450 enzymes. Genes from the phenylpropanoid pathway are cloned in the host under the control of the promoter, due to which secondary metabolites are often expressed heterologously. In an effort to promote flavonoids production, several promoters have been used depending on host requirements, including T7, ermE, and GAL1 promoters [41]. One of the limitations of microbiological flavonoids production was the extremely low concentration of malonyl-CoA. An increased production of flavonoids was achieved by co-expressing acetyl-CoA carboxylase genes from *Photorhabdus luminescens* [44]. Also essential for flavonoid biosynthesis is the presence of UDP-glucose. Using the udg gene, researchers knocked out the endogenous system for consuming UDP-glucose resulting in an increase in intracellular UDP-glucose concentrations and subsequently increased flavanones and anthocyanins production [45].

Scientists were able to generate a wider range of natural and unnatural products when combining bacteria and eukaryotic cells in a pot. Using a modified *S. cerevisiae* strain, de novo generation of the important flavonoid intermediate naringenin from glucose was achieved for the first time, leading to four times higher concentrations than those seen in previous de novo biosynthesis experiments [46, 47].

4. Chemistry of flavonoids

At the present scenario of scientific research, bioflavonoids are being considered as promising drug candidates, and extensive researches directed toward structural studies and biological efficacies of such class of compounds are in progress, which would eventually boost the on-going efforts leading to the discovery of new efficacious lead molecules. This remarkable class of natural compounds draws the attention of the scientists for their immense biological and pharmacological potentiality. Presently, over 10,000 individual flavonoid compounds are known, which are based on very few core structural skeletons (viz. flavone, flavanol, isoflavone, flavan, flavanone, chalcone, anthocyanin, coumarin etc. see in **Figure 1**) [9, 10, 30, 31].

Flavonoids are a group of natural compounds with low molecular weight polyphenolic substances based on the flavan nucleus are found mostly in plants. A novel chemical was extracted from oranges in 1930. It was given the name vitamin P at that time because it was thought to belong to a novel class of vitamins. Later, it was discovered that this material was a flavonoid (rutin), and as of now, more than 10,000 different flavonoid species have been found [48]. According to their chemical structure, flavonoids contain a 15-carbon skeleton. This skeleton consists of two benzene rings (A and B) linked by a heterocyclic pyrane ring (C) (**Figure 1**). These include flavones (such as luteolin apigenin and Itoside N), flavanols (such as kaempferol, myricetin and quercetin), flavanones (such as hesperetin, naringenin and abyssinoflavanone VI), and others. **Table 1** shows their general structures and other related information. Individual compounds within a class differ in the pattern of substitution of the A and B rings, whereas the many classes of flavonoids differ in the level of oxidation and pattern of substitution of the C ring [48]. Among the many forms of flavonoids, there are aglycones, glycosides, and methylated derivatives. Flavonoids contain aglycones as their basic structure (**Figure 1**). The α-pyrone (flavonols and flavanones) or its dihydro derivative (flavanols and flavanones) is a six-member ring that is condensed with the benzene ring (flavonols and flavanones) is a six-member ring that is condensed with the benzene ring. Based on the position of the benzenoid substituent, flavonoids are classified as flavonoids with a 2-position and isoflavonoids with a 3-position. Unlike flavanones, flavanols contain a hydroxyl group at the 3-position and a double bond

Class of flavonoids	Examples		
Flavones	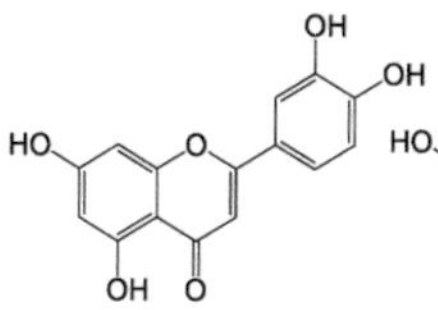 Luteolin	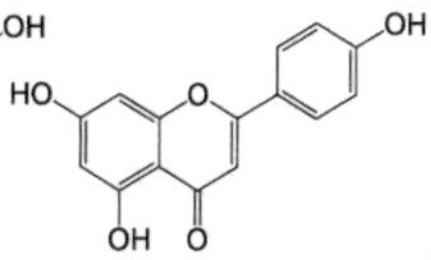 Apigenin	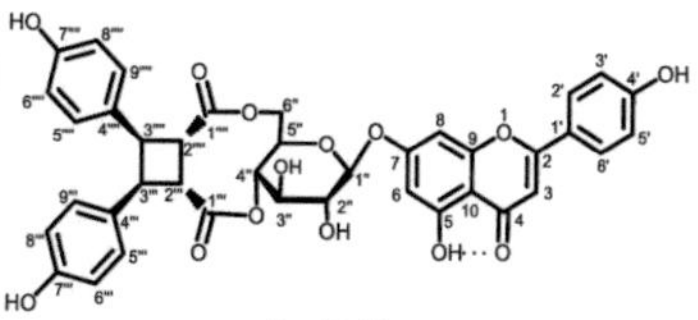 Itoside N
Flavans	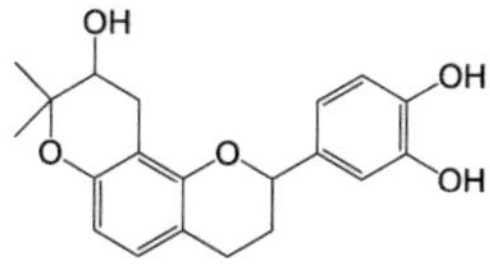 Brosimacutin K	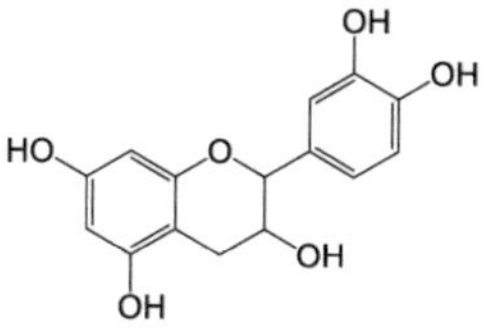 Catechin	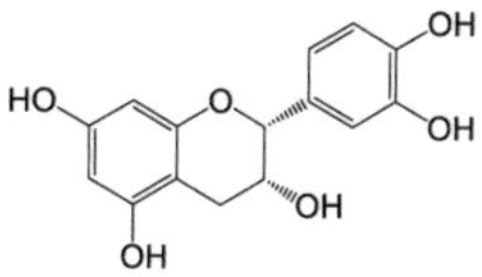 Epicatechin
Flavonols	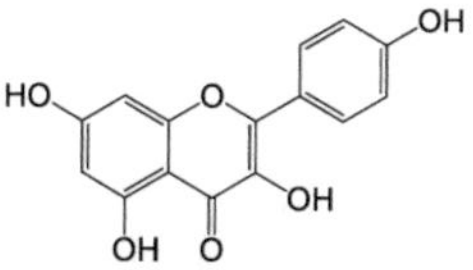 Kaempferol	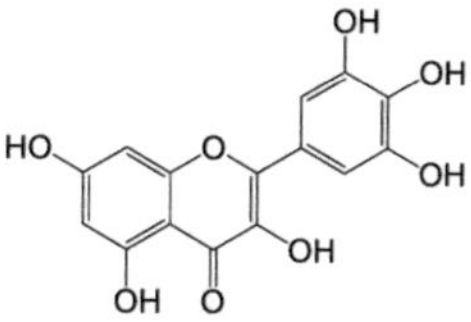 Myricetin	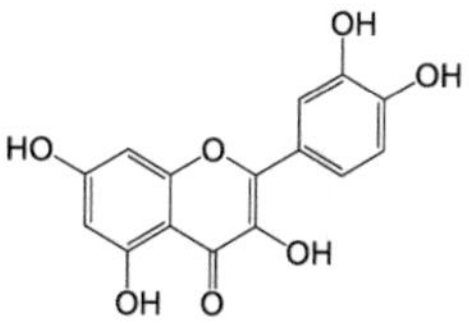 Quercetin
Isoflavones	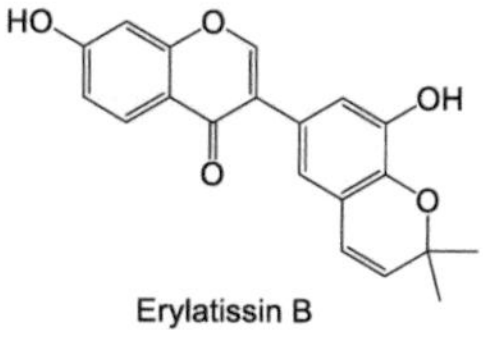 Erylatissin B	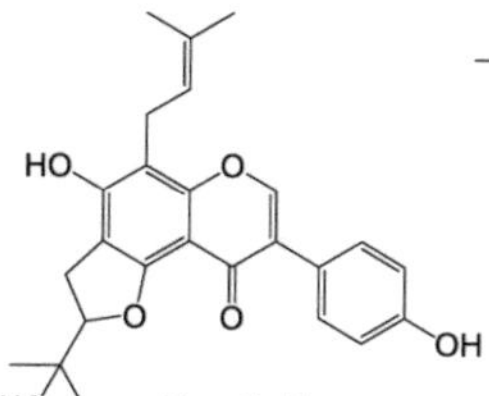 Vogelin H	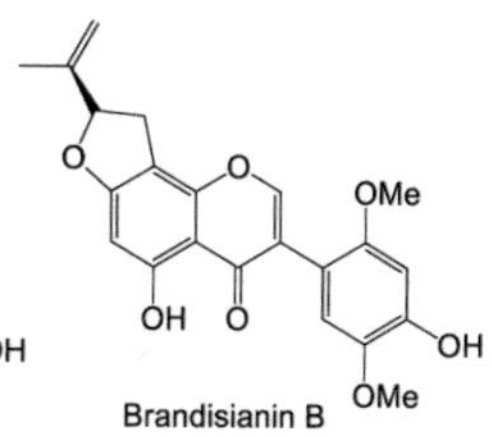 Brandisianin B
Isoflavanones	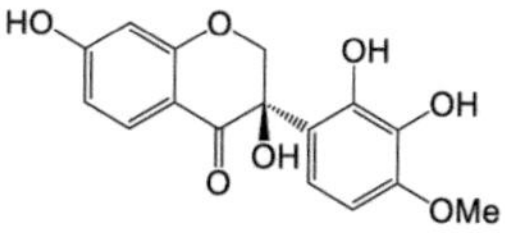 Isodarparvinol B	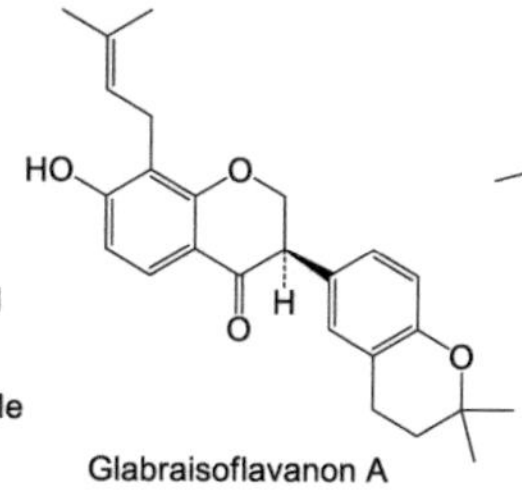 Glabraisoflavanon A	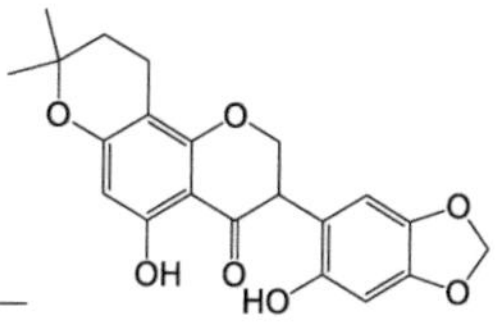 Discoloranone A
Flavanones	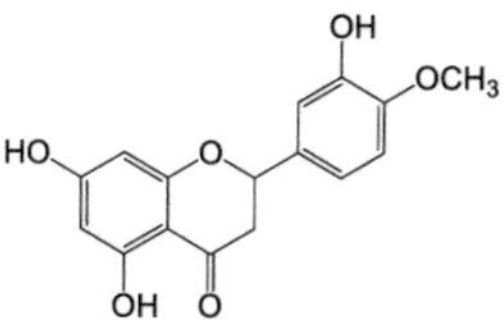 Hesperetin	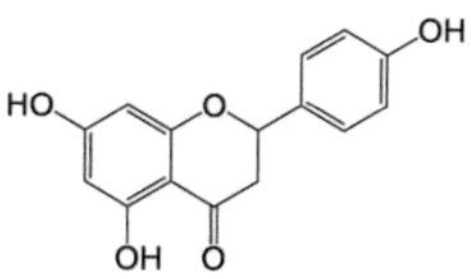 Naringenin	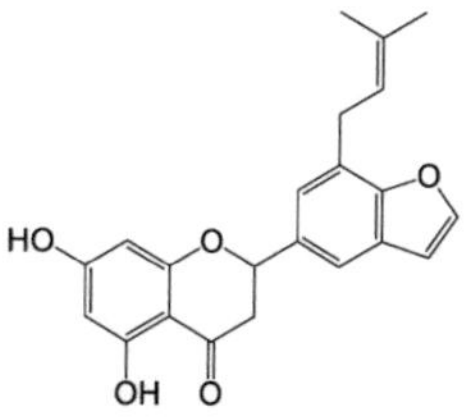 Abyssinoflavanone VI

Class of flavonoids	Examples		
Flavanols	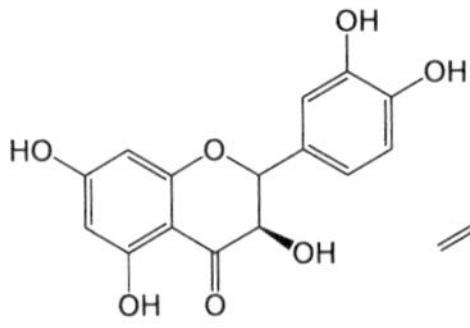 Taxifolin	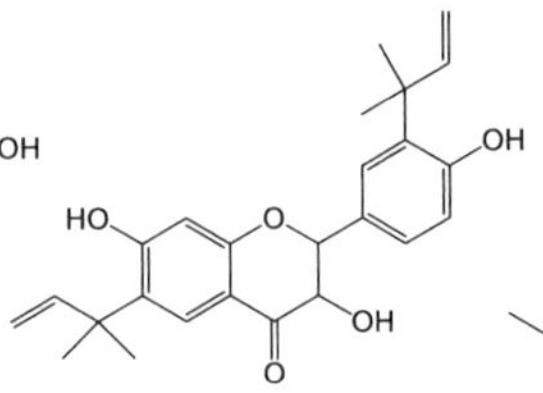 Lespeflorin B1	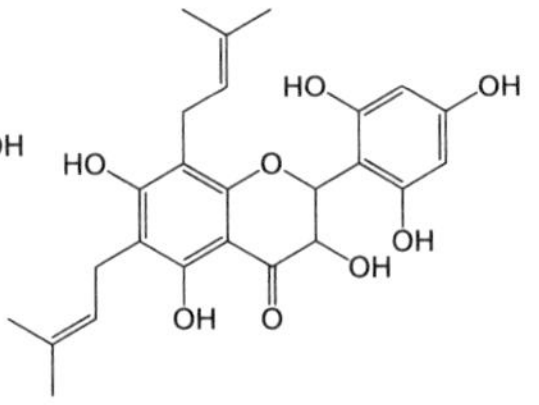 3-hydroxy-kenusanone B
Anthocyanins	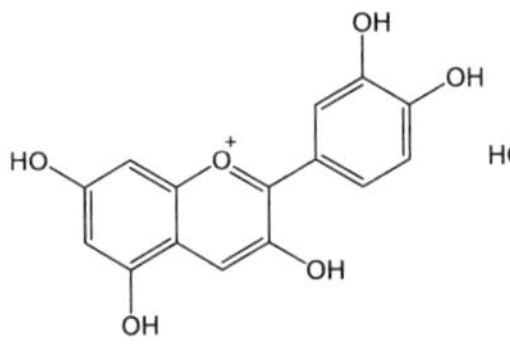 Cyanidin	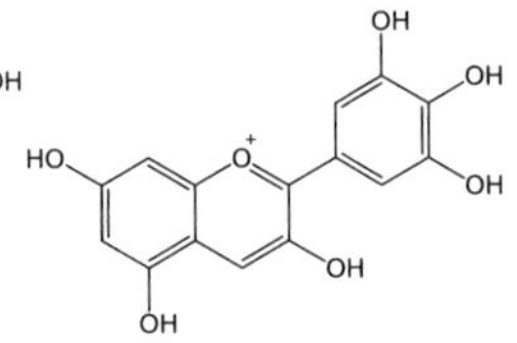 Delphinidin	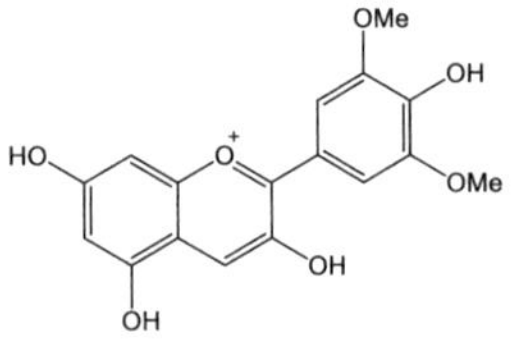 Malvidin
Chalcones	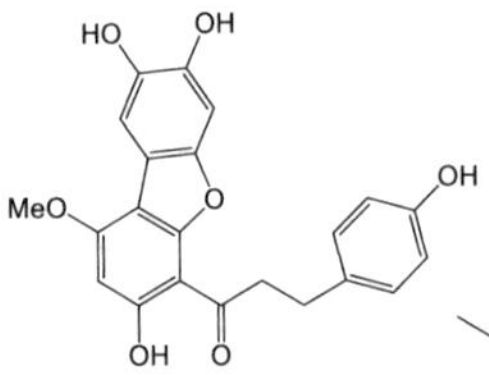 Pierotin B	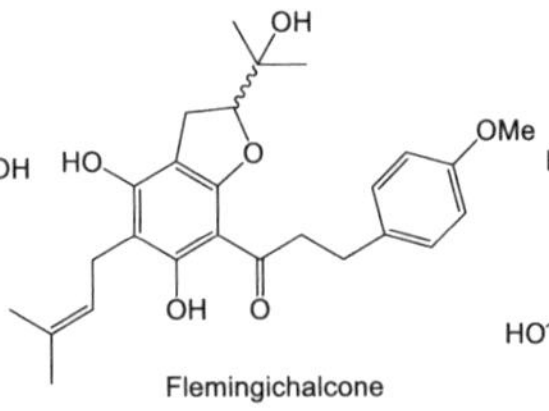 Flemingichalcone	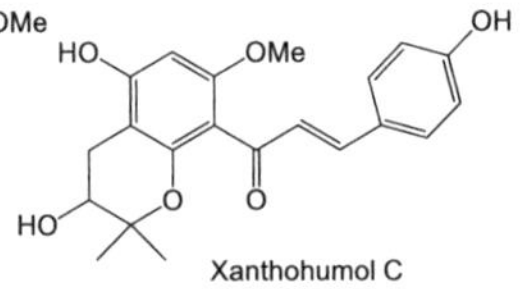 Xanthohumol C
Coumarins	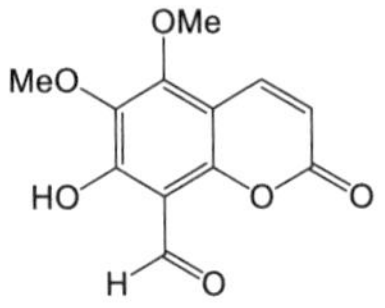 Yunngnin A	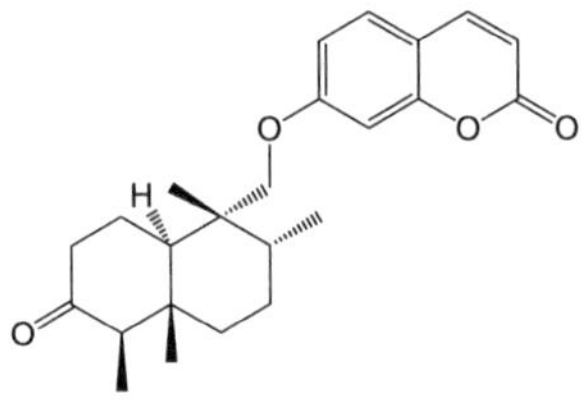 Szowitsiacoumarin A	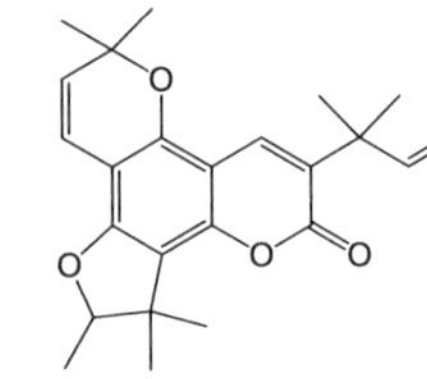 Citrumedin-B

Table 1.
Structure of some selective known flavonoids [9, 10, 49].

between C2 and C3 [49–51]. The most common positions of hydroxylation for flavonoids are 3, 5, 7, 2, 3′, 4′, and 5′. There is evidence to suggest that alcohol group methyl ethers and acetyl esters occur in nature. It is normally found that glycosides form when the glycosidic linkage appears in positions 3 or 7, and the carbohydrate is usually L-rhamnose, D-glucose, glucorhamnose, galactose, or arabinose [49, 50, 52].

4.1 Spectral characteristics of flavonoids

UV spectroscopic analysis of flavonoids identified two major absorption bands: Band I (320–385 nm) representing the absorption of the B ring, and Band II (250–285 nm) representing the absorption of the A ring. A shift in absorption can occur due to functional groups attached to flavonoid skeletons, such as 367 nm for kaempferol (3,5,7,4′-hydroxyl groups) and 371 nm for quercetin (3,5,7,3′,4′-hydroxyl groups) and 374 nm for myricetin (3,5,7,3′,4′,5′-hydroxyl groups) [53]. An absence of a 3-hydroxyl group distinguishes flavones from flavanols. According to their UV spectral properties, flavanones have a saturated heterocyclic C ring with no conjugation between the A and B rings [54]. Flavanones show only a shoulder for Band I at 326 and 327 nm and a very significant Band II absorption maximum between 270 and 295 nm, namely 288 nm for naringenin and 285 nm for taxifolin. In compounds with a monosubstituted B ring, Band II shows one peak (270 nm), but when a di-, tri-, or o-substituted B ring is present, it shows two peaks or one peak (258 nm) with a shoulder (272 nm). The color of anthocyanins varies with the quantity and position of the hydroxyl groups because they exhibit discrete Band I peaks in the 450–560 nm area due to the hydroxyl cinnamoyl system of the B ring and Band II peaks in the 240–280 nm region due to the benzoyl system of the A ring [55].

Nuclear magnetic resonance (NMR) spectroscopy has proven essential in the structural elucidation of natural products; it is one of the most effective methods available to natural product chemists [10].

In ^{1}H NMR investigations, the chemical shifts (δ) and the coupling constants (*J*), also known as and spin–spin couplings, are a good indicator. By comparing the recorded chemical shifts with the gathered data, this parameter provides important information on the relative number and kind of hydrogens. The number and anomeric configuration of the glycoside moieties connected to the aglycone, as well as the aglycone and acyl type groups associated to it, may all be determined using this [10]. The molecular architecture of flavone (Itoside N) can be learned a lot from the analysis of its 1H NMR spectrum data. The presence of an aromatic proton at C-3 in ring-C is shown by the one-proton singlet that appears at δ 6.86. The presence of two aromatic protons at C-6 and C-8, respectively, in ring-A is indicated by the signals that occurred as doublets (d) at δ 6.46 (1H, d, J = 2.0 Hz) and 6.80 (1H, d, J = 2.0 Hz). Four aromatic protons of the B-ring in the flavone skeleton may be the cause of the doublet (d) signals that emerged at δ 7.97 (2H, dd, J = 8.0 Hz) for two protons and δ 6.97 (2H, dd, J = 8.0 Hz) for another pair of protons. Ring B is definitely para-disubstituted, according to the chemical shifts and coupling constant values for its four protons. Moreover, the ^{1}H NMR spectrum of Itoside N (**Table 1**) indicates that a partial structure similar in structure to p,p′-dihydroxy-μ-truxinic acid in Itoside N is formed by two p-dihydroxy benzenoid groups that combination with a cyclobutane [δ 42.9 (C-2‴), 43.5 (C-3‴), 45.5 (C-2⁗), 42.9 (C-3⁗)] moiety [9, 10].

When combined with ^{1}H NMR data, ^{13}C NMR data can be used to determine the types of groups that are present in molecules. It should be noted, however, that ^{13}C NMR is not as responsive as ^{1}H NMR because ^{13}C is less abundant (1.1%) than ^{1}H (99.9%) [10]. The C-2/C-II-2 and C-3/C-II-3 sp2-hybridized carbons can be found, respectively, at δ_C 152.5–165.5 and 103–132.1 in ^{13}C-NMR spectra. According to ^{13}C NMR data of flavonoid compounds, C-4/C-II-4 appear between δ_C 176.2 and δ_C 182.9 when C-2-C-3 is unsaturated (sp2), but when C-2/C-3 is sp3 hybridized, a down-field shift of C-4/C-I-4 is typically observed between δ_C 196.2 to δ_C 197.9. The typical range of δ_C 159.6–164.7 includes the aromatic C-5, C-6, C-7, C-8, C-9/C-4a, and C-10/C-8a. Generally CMR of glucopyranosyl moiety appeared in the range of δ_C 60.6–102.11; rhamnopyranosyl (Rha) also appeared in the range of δ_C 17.23–101.0; and glucuronopyranoside showed the value in δ_C 98.4–171.6. The value of glucuronopyranoside is nearly identical to that of the glucopyranosyl moiety, however the carboxylic group is what gives the compound its high value (δ_C 171.6). The flavone, Itoside N is found to appear around a range δ_C of 42.9–172.2 for the 4″/6″-p-hydroxy-μ-truxinyl group [10].

4.2 Substitution pattern of flavonoids

The main flavonoid classes and some of their discovered structural variants are shown in **Table 2**. Within the primary classes, flavonoids' structures differ significantly by substitutions such as hydroxylation, glycosylation, hydrogenation, methylation, malonylation, and sulphation etc. Many flavonoids are found in nature

Name of flavonoids	No. of hydroxyl (-OH) groups	Position of OH groups	Other substitutions on the basic structure	Position of the substitutions
Myricetin	6	3, 5, 7, 3′, 4′, 5′	—	—
Gossypetin	6	3, 5, 7, 8, 3′, 4′	—	—
Quercetagen	6	3, 5, 6, 7, 3′, 4″	—	—
Hypolactin	5	5, 7, 8, 3′, 4′	—	—
Quercetin	5	3, 5, 7, 3′, 4′	—	—
Myricetrin	5	5, 7, 3′, 4′, 5′	O-Rha	3
Rutin	4	5, 7, 3′, 4′	O-Rut	3
Kaempferol	4	3, 5, 7, 4′	—	—
Quercetrin	4	5, 7, 3′, 4′	O-Rha	3
Fisetin	4	3, 7, 3′, 4′	—	—
Rhamnetin	4	3, 5, 3′, 4′	O-Me	7
Orientin	4	5, 7, 3′, 4′	Glc	8
Apigenin	3	5, 7, 4′	—	—
Galangin	3	3, 5, 7	—	—
Kaempferide	3	3, 5, 7	O-Me	4′
Luteolin-7-glucoside	3	5, 3′, 4′	O-Glc	7
Vicenin-2	3	5, 7, 4′	Glc	6, 8
Sideritoflavone	3	5, 3′, 4′	O-Me	6, 7, 8
Pinocembrin	2	5, 7		
Gardenin-D	2	5, 3′	O-Me	6, 7, 8, 4′
Diosrnin	2	3, 3′	O-Rut, O-Me	5, 4′
Robinin	2	5, 4′	O-Galc-Rha, Rha	3, 7′
Cirsimaritin	2	5, 4′	O-Me	6, 7
Xanthomicrol	2	5, 4′	O-Me	6, 7, 8
8-Methoxycirisilincol	2	5, 4′	O-Me	6, 7, 8, 3′
3-OH-Flavone	1	3		—
Techtochyrsin	1	5	O-Me	7
Troxerutin	1	5	O-Rut, O-He, O-He, O-He	3, 7, 3′, 4′

Table 2.
Some example of substitution pattern of flavonoids [9, 49, 59].

as flavonoid glycosides, and D-glucose, L-rhamnose, glucorhamnose, galactose, lignin, and arabinose are some examples of carbohydrate substitutes [9, 49, 50]. The most prevalent flavonoid glycosides in the diet are quercitrin, rutin, and robinin. Intestinal flora hydrolyzes them to create the physiologically active aglycone (sugar-free flavonoid). Due to its prominence as the primary flavonoid present in foods, quercetin has been the focus of numerous studies examining the biological impacts of flavonoids [9, 49, 50].

4.3 Polymerization of flavonoids

In term of units of flavonoids molecules there are three types of flavonoids namely monomers, dimers, and oligomers. There are huge differences between the molecular weights of different monomers. Polymers of flavonoids make up condensed tannins. Epicatechin, epigallocatechin, epicatechin gallate, and epigallocatechin gallate are the four primary catechin components found in tea tannins. The main catechin in tea, epigallocatechin gallate, accounts for more than half of the total catechin content. The dimeric theaflavins and polymeric thearubigins of black Indian tea, which generate brightness and astringency, respectively, are produced by enzymatic oxidation of tea catechins during fermentation of macerated tea leaves [56, 57]. Thearubigins come in a wide variety of sizes, from molecules with up to 100 flavonoid units to oligomers of four or five units [56]. Green "Chinese" tea does not undergo fermentation during processing, in contrast to black tea, hence its flavonoids largely exist as monomers. The anthocyanins and other flavonoids in red wine polymerize to create tannins, which give the wine its distinctive hues, tastes, and astringency [58, 59].

5. Role of flavonoids in plants

Plants are the key source of natural products and plants had already yielded a vast number of phytochemicals and still continue to a major source of biologically active molecules. Numerous plants have already established their potentiality as a source of naturally occurring insecticides, pesticides, fungicides and agro-chemicals as an alternative to toxic and hazardous synthetic chemicals. Owing to ever-increasing awareness to the hazardous side effects of synthetic chemicals, more and more emphasis is being given on the use of products obtained from natural sources so that ecological balance is well-maintained. The WHO has already called for an immediate ban on the use of many synthetic chemicals viz. endosulfan is a dangerous synthetic pesticide that causes severe damage to the eyes, kidneys, and liver. The Government of India had already banned the use of 12 highly toxic and hazardous pesticides and imposed restriction on the use of many others to prevent environmental pollution. To minimize the hazardous effects and to control environmental pollution, attempts are now being made to develop naturally occurring plant-based pesticides. Many phyto-chemicals such as phytoecdysones, and azadirachtin (from Indian neem) have been reported to possess pesticidal and insecticidal properties and are being widely used in protecting the loss of crop from the attack of insects and parasites in place of their synthetic analogues because of their non-toxic, non-pollutant, readily bio-degradable character and harmless nature of their residues.

As a result of changes in plant growth, conditions, and maturity, there are more than 10,000 types of flavonoid compounds found in vascular plants, which vary in type and quantity according to a variety of factors. There is still a lack of systematic analysis of the flavonoid content of many plant species, which makes it difficult to identify and quantify all the flavonoids humans consume [57]. In order to defend themselves against herbivores, pathogens, oxidative cell damage, and fungal parasites, plants have evolved to synthesize flavonoids [58]. On the other hand, flavonoids act as a stimulant that aids in pollination and guides insects on their way to food sources. For example, flavonoid compounds anthocyanins are responsible for the pink, blue red, light purple and violet colors in flowers, fruits, and vegetables [56, 59].

5.1 Combating oxidative stress of flavonoids

It has long been reported that flavonoids have a variety of functions in plants [60]. Both abiotic and biotic factors contribute to oxidative stress in plants as a result of ROS being generated in plants. A high level of oxidative stress commonly enhances the synthesis of flavonoids in plants. The pigments are capable of absorbing the most energetic rays of the sun (i.e., UV-B and UV-A), inhibiting ROS production, and quenching ROS once they have been generated [61]. When plants moved from water to soil, flavonoids were primarily responsible for screening UV-B. Different flavonoids have different antioxidant capacities and UV-wavelength-absorbing capabilities based on their substitutions. There is an increase in antioxidant capacity in flavonoids with dihydroxy B rings substituted for these rings, whereas flavonoids with monohydroxy B rings have a greater ability to absorb UV wave lengths. Glycosylation is generally the hallmark of the most reactive hydroxyl groups of flavonoids (7-OH in flavones or 3-OH in flavanols). Flavonoids can be transported from the endoplasmic reticulum to various cellular compartments and secreted from their plasma membrane and cell wall through glycosylation, thereby increasing their solubility in the aqueous cellular environment, protecting the reactive hydroxyl groups from auto-oxidation [62]. Studies have shown that antioxidant flavonoids are found in cells of the mesophyll and in chloroplasts, which generate ROS. Using this method, they are able to easily quench H_2O_2, hydroxyl radicals, and singlet oxygen [61, 63]. Conditions that restrict CO_2 diffusion to carboxylation sites and carboxylation efficiency may exacerbate oxidative stress caused by an excessive amount of excitation energy in chloroplasts [61, 64]. A number of environmental factors can restrict CO_2 assimilation, including drought/salinity, temperature fluctuations, and nutrient scarcity. This can decrease the activity of ROS detoxifying enzymes in the chloroplast [65], which increases the production of antioxidant flavonoids. Flavonoids are highly important for plants under severe stress conditions because of their reducing properties. In addition to their functional roles, dihydroxy B ring substitutes are also highly concentrated [66]. It has been suggested that flavonoids represent a secondary antioxidant defense system in plants under stress [61]. In response to oxidative stress, lipid peroxidation occurs, causing cell membrane degradation. It has been suggested that quercetin-3-*O*-rutinoside (Rutin) interacts with phospholipids polar head at the water lipid interface, increasing membrane inflexibility and thereby protecting membranes from oxidative damage [67].

5.2 Role of flavonoids as growth regulator

Role of Flavonoids as Growth Regulator: in plant-environment interactions, flavonoids play an essential role. There is evidence that flavonoids control auxin catabolism and movement by using in nanomolar range. When flavonoids produce auxin gradients, they produce phenotypes with different morphoanatomical characteristics [68]. Stress-induced morphogenic responses of plants are controlled largely by flavonoids, which may have a direct relevance to flight strategies of sessile organisms exposed to unfavorable environments [69]. A species that produces dihydroxy flavonoids exhibits its phenotypic characteristics that are strikingly different from a species that produces monohydroxy flavonoids [70]. In sunny situations, dwarf bushy phenotypes with few, tiny, and thick leaves are typically prevalent, shielding leaves deep in the canopy from light-induced severe cellular homeostasis disruptions. Alternatively, shaded plants, which contain kaempferol and/or apigenin derivatives, have long internodes and large leaf lamina, along with reduced leaf thickness [69]. PIN/MDR glycoproteins that facilitate cell-to-cell movement of auxin are inhibited by flavonoids at the plasma

Class of flavonoids	Name of flavonoids	Dietary sources
Anthocyanidin	Peonidin	Cranberries, blueberries, plums, grapes, cherries, sweet potatoes
Catechin	Theaflavin	Tea leaves, black tea, oolong tea
Coumarin	Scopoletin	Vinegar, dandelion coffee
Flavan	Epicatechin	Milk, chocolate, commercial, reduced fat
Flavanol	Taxifolin	Vinegar, citrus fruits
Flavanone	Abyssinones	French bean seeds
	Eriodictyol	Lemons, rosehips
	Hesperidin	Bitter orange, petit grain, orange, orange juice, lemon, lime
	Naringenin	Grapes
Flavone	Diosmetin	Vetch
	Apigenin	Milk, chocolate, commercial, reduced fat
	Luteolin	Celery, broccoli, green pepper, parsley, thyme, dandelion, perilla, chamomile tea, carrots, olive oil, peppermint, rosemary, navel oranges, oregano
	Tricin	Rice bran
Flavanol	Fisetin	Strawberries, apples, persimmons, onions, cucumbers
	Kaempferol	Apples, grapes, tomatoes, green tea, potatoes, onions, broccoli, Brussels sprouts, squash, cucumbers, lettuce, green beans, peaches, blackberries, raspberries, spinach
	Myricetin	Vegetables, fruits, nuts, berries, tea, red wine
	Rutin	Green tea, grape seeds, red pepper, apple, citrus fruits, berries, peaches
	Quercetin	Vegetables, fruits and beverages, spices, soups, fruit juices
Isoflavone	Biochanin	Red clover, soya, alfalfa sprouts, peanuts, chickpeas (Cicer arietinum), other legumes
	Daidzein	Soyabeans, tofu
	Genistein	Fats, oils, beef, red clover, soyabeans, psoralea, lupin, fava beans, kudzu, psoralea

Table 3.
Example of some flavonoids and their rich dietary sources [74].

membrane. In flavonoids, the catechol group is present at the B ring of the flavonoid skeleton that is responsible for inhibiting the activity of the efflux facilitator PIN and MDR proteins. The chemical structure of flavonoids also influences their action on IAA-oxidase significantly [71]. In recent years, it has been found that flavonoids can influence the activity of proteins involved in cell growth due to a nuclear location of flavonoids as well as the actions of enzymes that produce flavonoids [72]. This suggests that flavonoids may be able to regulate transcription [73]. In **Table 3** showed some flavonoids and their rich dietary sources [74].

6. Role of flavonoids in pest control

There is an ever-growing demand for natural pesticides from plants. As an alternative to synthetic pesticides, flavonoids are being used to develop new pesticides. A variety of insect larvae can be prevented from growing if they are inhibited by these compounds [75]. It is known that some flavonoids inhibit the production of juvenile hormone which is involved in molting and reproduction in several insects [76]. A number of flavonoids have been shown to suppress agricultural pest activity, such as oviposition, fecundity, mortality, weight reduction, and the emergence of adults [77, 78]. In their article, Lena Schnarr et al. [79] reported 281 different pesticidal active flavonoids that were investigated in either pure form or as extracts containing flavonoid, with the most studied compounds being quercetin, kaempferol, apigenin, luteolin and their glycosides [79]. In another study, Quercetin, rutin, and naringin were all effective in controlling *Eriosoma lanigerum* Hausmann nymphs and adults. An integrated management program for this aphid can use these products as an insecticide [80]. Flavonoids may have an insecticidal effect depending on their concentration; if too low, they are ineffective [81]; as a result, it is crucial to determine the minimum concentration for flavonoids to be effective [79, 81].

7. Patent information

About 40 patent information on agriculture related matter including all necessary agenda are presented in **Table 4**, which deal with, synthesis of various flavonoids in plant and their analogs, method for increasing the flavonoids content, pest control and pesticidal activity of plant and plant protection.

8. Concluding remarks

Naturally occurring flavonoids are of much interest to the scientific community at a large due to their multidirectional therapeutic applications. Besides, knowledge about natural distribution of flavonoids of varying structural skeletons is also very much essential to the taxonomists for classifying plants in the light of chemotaxonomy. Thus, flavonoids are of much interest to the workers of interdisciplinary fields.

The use of plant flavonoids could provide eco-friendly and sustainable approaches to improving food quality and crop yield as well as improving their adaptation to environmental stress. When applied in practice, flavonoids could be very effective in the field, as a result of their phytotoxic and pesticidal properties. Also natural

Patent number	Filing date	Issue date	Original assignee	Title	Inventors	Refs
US2465854A	26-08-44	29-03-49	Shell Dev	Insecticidal composition containing an aromatic unsaturated carbonyl compound	S.C. Dorman, S.A. Ballard	[82]
US4193984A	09-04-76	18-03-80	Herculite Protective Fabrics Corporation	Method and compositions for controlling flying insects	A.F. Kydonieus	[83]
JPS57120501A	19-01-81	27-07-82	Kiyoshi Saotome	Protecting method of field crop	K. Saotome	[84]
FR2529755A1	06-07-82	13-01-84	Kiyoshi Saotome	Crop protection method by means of Cinnamaldehyde	K. Saotome	[85]
JPH01261303A	13-04-88	18-10-89	Taiyo Koryo Kk	Insect pest repellent	H. Miyawaki, K. Saotome	[86]
US5149715A	09-02-89	22-09-92	Monterey Mushroom, Inc.	Control of fungal diseases in the production of mushrooms	G.L. Armstrong, N.S. Dunn-Coleman, M. Wach	[87]
WO1996020594A1	29-12-95	11-07-96	Proguard Inc. (US)	Use of flavonoid aldehydes as insecticides	R.W. Emerson, B.G. Crandall, Jr	[88]
US6548085B1	30-03-99	15-04-03	Woodstream Corp	Insecticidal compositions and method of controlling insect pests using same	K.A. Zobitne, M.J. Gehret	[89]
US6593299B1	02-08-99	15-07-03	University of Florida Research Foundation Inc. Insect Biotechnology Inc	Compositions and methods for controlling pests	J. Bennett, A. Brandt, D. Borovsky	[90]
US6841577B2	18-12-01	11-01-05	GIBRALTAR BUSINESS CAPITAL LLC Kittrich	Pesticidal activity of plant essential oils and their constituents	S.M. Bessette, M.A. Beigler	[91]
US20050031761A1	07-05-04	10-02-05	SUNRISE COFFEE CO	Methods of producing a functionalized coffee	D. Brucker, M. Sweeney, T. Breen	[92]
US20050208643A1 US7604968B2	01-03-05	22-09-05 20-10-09	University of Minnesota	Microorganisms for the recombinant production of resveratrol and other flavonoids	C. Schmidt-Dannert, K. Watts	[93]

Patent number	Filing date	Issue date	Original assignee	Title	Inventors	Refs
US20070232495A1	23-03-07	04-10-07	Nappa Alvaro O, Lorenzini Felipe C, Sanhueza Andres L	Compositions and methods to add value to plant products, increasing the commercial quality, resistance to external factors and polyphenol content thereof	A. Nappa, F. Lorenzini, A. Sanhueza	[94]
US20060019334A1	11-07-05	04-03-08	Research Foundation of State University of New York	Production of flavonoids by recombinant microorganisms	M. Koffas, E. Leonard, Y. Yan, J. Chemler	[95]
US20080274519A1	07-07-08	06-11-08	BIORES HEALTH Ltd	Flavonoid concentrates	R.G. Wallace	[96]
US7750211B2	10-09-03	06-07-10	Noble Research Institute LLC	Methods and compositions for production of flavonoid and isoflavonoid nutraceuticals	R.A. Dixon, C.-J. Liu, B. Deavours	[97]
NZ580217A	23-04-08	29-07-11	Samuel Roberts Noble Found Inc	Production of proanthocyanidins to improve forage quality	R.A. Dixon, L.V. Modolo, G. Peel	[98]
US8142801B2	02-02-10	27-03-12	HOMS LLC	Pesticidal compositions and methods of use thereof	A. Jones	[99]
AU2010306410A1 AU2010306410B2	15-10-10	03-05-12 13-08-15	Agriculture Victoria Services Pty Ltd	Manipulation of flavonoid biosynthetic pathway	A. Mouradov, G. Spangenberg	[100]
JP5002848B2	13-04-10	15-08-12	Suntory Holdings Ltd	Flavonoid 3′,5′ hydroxylase gene sequence and method of use thereof	P. B. Yoshikazu, T. J. Mason	[101]
MX2011010032A	23-09-11	25-03-13	M.S. Aguilar, Y.M. H. Romero, C.M.R. Narvaez	Pesticide made of isoquinoline alkaloids, flavonoids and vegetable and/or essential oils.	M.S. Aguilar, Y.M. H. Romero, C.M.R. Narvaez	[102]
US20130340118A1 US9567600B2	26-08-13	19-12-13 14-02-17	Agriculture Victoria Services Pty Ltd	Modification of flavonoid biosynthesis in plants	A. Mouradov, G. Spangenberg	[103]

Patent number	Filing date	Issue date	Original assignee	Title	Inventors	Refs
WO2014122446A1	05-02-14	14-08-14	Phyto Innovative Product Ltd.,	Plant protection composition and method	I. Ripley	[104]
US8877219B2	31-01-13	04-11-14	Kittrich Corp	Pesticidal compositions containing rosemary oil and wintergreen oil	S.M. Bessette, A. D. Lindsay	[105]
US 2014/0335210A1	13-05-13	13-11-14	FLAVITPURE, INC., Cheyenne	Method and agrochemical composition for using larch wood extracts in agriculture	S.V. Philippov, I. M. Bogorodov	[106]
AU2013338110A1 AU2013338110B2	29-10-13	30-04-15 01-12-16	Novozymes BioAg AS	Compositions and methods for enhancing plant growth	L. Blankenshi, A. Habib, Y. Kang, S. Semones	[107]
CN104640461A CN104640461B	06-08-13	20-05-15 19-12-17	Nestec SA	Anthocyanin colored composition	N. Gallifer, M. Meeker, S. Carwin, K. Boltlick, P. Choisy	[108]
US 2015/0216181A1	21-03-12	06-08-15	PROMOTORATECNICA INDUSTRIAL, S.A. DE C.V., Jiutepec, Estado de Mexico	Pesticide having an insecticide, acarcide and nematicde action based on isoquinoline alkaloids and flavonoids	Y.M.H. Romero, C.M.R. Narvaez, M.S. Aguilar	[109]
CN104823979A	18-05-15	12-08-15	South China Agricultural University	Application of flavonoid compound theaflavanoside II to prevention and treatment of plant nematode diseases	W. Yanhua, S.Y. Liao, J. X. Hui	[110]
EP2906055A1 EP2906055B1	08-10-13	19-08-15 16-12-20	RJ Reynolds Tobacco Co	Method of extracting tobacco-derived o-methylated flavonoid and use thereof	A. R. Gerardi	[111]
ES2464642B1	21-03-12	20-11-15	PROMOTORA TECNICA INDUSTRIAL	Pesticide with insecticide, acaricide and nematiciated action based on isoquinolinic alcaloids and flavonoids	H. Romero, R. Narvaez, S. Aguilar	[112]

Patent number	Filing date	Issue date	Original assignee	Title	Inventors	Refs
US9439886B2	27-04-15	13-09-16	Agency for Science Technology and Research Singapore	Methods for producing crosslinked flavonoid hydrogels	M. Kurisawa, F. Lee, J. E. Chung, P.Y. P. Chan	[113]
US9523089B2	13-09-13	20-12-16	Agriculture Victoria Services Pty Ltd	Manipulation of flavonoid biosynthesis in plants	G. Spangenberg, T.I. Sawbridge, E.-K. Ong, M. Emmerling	[114]
US9580725B2	26-06-14	28-02-17	Norfolk Plant Sciences Ltd	Methods and compositions for modifying plant flavonoid composition and disease resistance	J. Luo, E. Butelli, J. Jones, L. Tomlinson, C.R. Martin	[115]
KR101802249B1	03-08-15	29-11-17	Sunjin Industry Co., Ltd. Eco Farm	Natural composition and method for manufacturing the composition avoiding and/or controlling the Hemiptera	K. In-Gyu, K. Chang-Gil, K. Yeol, L. Mok-Hyeong, Y. Hae, G. Jeong, H.Y. Choi, C. Han-Jeung, J. Jae-Doo, Y. Hwan-Sang	[116]
US9949490B2	02-07-14	24-04-18	Ralco Nutrition, Inc.	Agricultural compositions and applications utilizing essential oils	R.D. Lamb, M.D. Johnson	[117]
WO2019016806A1	19-07-17	24-01-19	Future Tense Technological Development and Entrepreneurship Ltd	Pesticide containing antioxidants	Y. Tsivion	[118]
US10285393B2	24-10-14	14-04-19	Red Band Traps, Llc	Arthropod pest trapping device, system and method	T.V. Bailey	[119]
RU2729743C1	27-02-20	11-08-20	Federal State Budgetary Scientific Institution	Method for increasing content of flavonoids in buckwheat fruits	A.G. Klykov, G.A. Murugova, O.A. Timoshinova, S.A. Borovaya, E.L. Chaikina	[120]
EP3912470A1	16-01-20	24-11-21	SDS Biotech Corp Idemitsu Kosan Co Ltd	Plant growth regulating agent	S. Ishida, K. Inai, M. Tanaka, T. Nomoto	[121]

Table 4.
Some patent information on agriculture related matter.

herbicides made from bioflavonoids are being investigated more and more in integrated weed control.

Further research and investigations are required to understand the full range of activity of flavonoids produced naturally and/or applied artificially for batter benefit in the field of agriculture.

Acknowledgements

SKJ is grateful to the Department of Chemistry, Krishna Chandra College, Hetampur for providing necessary infrastructural facilities to carry out this work.

Conflict of interest

The authors declare no conflict of interest.

Author details

Shyamal K. Jash
Department of Chemistry, Krishna Chandra College (Affiliated to the University of Burdwan), Birbhum, West Bengal, India

*Address all correspondence to: jash_sh@yahoo.co.in

References

[1] Harborne JB, Williams CA. Anthocyanins and other flavonoids. Natural Product Reports. 2001;**18**:310-333

[2] Williams CA, Grayer RJ. Anthocyanins and other flavonoids. Natural Product Reports. 2004;**21**:539-573

[3] Reynaud J, Guilet D, Terreux R, Lussignol M, Walchshofer N. Isoflavonoids in non-leguminous families: An update. Natural Product Reports. 2005;**22**:504-515

[4] Mackova Z, Koblovska R, Lapcik O. Distribution of isoflavonoids in non-leguminous taxa – an update. Phytochemistry. 2006;**67**:849-855

[5] Andersen ØM, Markham KR, editors. Flavonoids: Chemistry, Biochemistry and Applications. 1st ed. Boca Raton: CRC Press; 2006

[6] Li J, Jiang Y. Litchi Flavonoids: Isolation, Identification and Biological Activity. Molecules. 2007;**12**:745-758

[7] Veitch NC. Isoflavonoids of the Leguminosae. Natural Product Reports. 2007;**24**:417-464

[8] Veitch NC, Grayer RJ. Flavonoids and their glycosides, including anthocyanins. Natural Product Reports. 2008;**25**:555-611

[9] Jash SK, Brahmachari G. Recent progress in the research of naturally occurring flavonoids: A look through. Signpost Open Access Journal of Organic and Biomolecular Chemistry. 2013;**1**:65-168

[10] Jash SK, Gorai D, Mandal LC, Roy R. Nuclear magnetic resonance spectroscopic behaviour of some selective natural flavonoids: A look through. Mini-Reviews in Organic Chemistry. 2020;**17**(2):185-196

[11] Liu W, Feng Y, Yu S, Fan Z, Li X, Li J, et al. The flavonoid biosynthesis network in plants. International Journal of Molecular Sciences. 2021;**22**:12824

[12] Observatory Earth. World of change: Global temperatures [Internet]. 2020. Available from: https://earthobservatory.nasa.gov/world-of-change/global-temperatures. [Accessed: 24 March 2020]

[13] Richardson Y, Blin J, Julbe A. A short overview on purification and conditioning of syngas produced by biomass gasification: Catalytic strategies, process intensification and new concepts. Progress in Energy and Combustion Science. 2012;**38**:765-781

[14] EPA. Global emissions by gas [Internet]. 2020. Available from: https://www.epa.gov/ghgemissions/global-greenhousegas-emissions-data. [Accessed: 24 March 2020]

[15] Scialabba NE-H, Müller-Lindenlauf M. Organic agriculture and climate change. Renewable Agriculture and Food Systems. 2010;**25**:158-169

[16] Chalker-Scott L. Environmental significance of anthocyanins in plant stress responses. Photochemistry and Photobiology. 1999;**70**:1-9

[17] Chen S, Wu F, Li Y, Qian Y, Pan X, Li F, et al. NtMYB4 and NtCHS1 are critical factors in the regulation of flavonoid biosynthesis and are involved in salinity responsiveness. Frontiers in Plant Science. 2019;**10**:1-13

[18] Shojaie B, Mostajeran A, Ghanadian M. Flavonoid dynamic

responses to different drought conditions: Amount, type, and localization of flavonols in roots and shoots of Arabidopsis thaliana L. Turkish Journal of Biology. 2016;**40**:612-622

[19] Brunetti C, Di Ferdinando M, Fini A, Pollastri S, Tattini M. Flavonoids as antioxidants and developmental regulators: Relative significance in plants and humans. International Journal of Molecular Sciences. 2013;**14**:3540-3555

[20] Gori A, Tattini M, Centritto M, Ferrini F, Marino G, Mori J, et al. Seasonal and daily variations in primary and secondary metabolism of three maquis shrubs unveil different adaptive responses to mediterranean climate. Conservation Physiology. 2019;7:coz070

[21] Gori A, Nascimento LB, Ferrini F, Centritto M, Brunetti C. Seasonal and Diurnal Variation in Leaf Phenolics of Three Medicinal Mediterranean Wild Species: What Is the Best Harvesting Moment to Obtain the Richest and the Most Antioxidant Extracts? Molecules. 2020;**25**:956

[22] Grotewold E. The genetics and biochemistry of floral pigments. Annual Review of Plant Biology. 2006;**57**:761-780

[23] Singla P, Garg N. Plant flavonoids: Key players in signaling, establishment, and regulation of rhizobial and mycorrhizal endosymbioses. In: Mycorrhiza-Function, Diversity, State of the Art. Berlin/Heidelberg, Germany: Springer; 2017. pp. 133-176

[24] Liu CW, Murray JD. The role of flavonoids in nodulation host-range specificity: An update. Plants (Basel). 2016;**5**:33

[25] Mierziak J, Kostyn K, Kulma A. Flavonoids as important molecules of plant interactions with the environment. Molecules. 2014;**19**:16240-16265

[26] Dudek B, Warskulat A-C, Schneider B. The Occurrence of Flavonoids and Related Compounds in Flower Sections of Papaver nudicaule. Plants. 2016;**5**:28

[27] Cesco S, Mimmo T, Tonon G, Tomasi N, Pinton R, Terzano R, et al. Plant-borne flavonoids released into the rhizosphere: Impact on soil bio-activities related to plant nutrition. A review. Biology and Fertility of Soils. 2012;**48**:123-149

[28] Dong NQ, Lin HX. Contribution of phenylpropanoid metabolism to plant development and plant-environment interactions. Journal of Integrative Plant Biology. 2020;**63**:180-209

[29] Nabavi SM, Šamec D, Tomczyk M, Milella L, Russo D, Habtemariam S, et al. Flavonoid biosynthetic pathways in plants: Versatile targets for metabolic engineering. Biotechnology Advances. 2020;**38**:107316

[30] Sasaki N, Nakayama T. Achievements and Perspectives in Biochemistry Concerning Anthocyanin Modification for Blue Flower Coloration. Plant & Cell Physiology. 2015;**56**:28-40

[31] Flavonoid W-SB, Biosynthesis. A Colorful Model for Genetics, Biochemistry, Cell Biology, and Biotechnology. Plant Physiology. 2001;**126**:485-493

[32] Samanta A, Das G, Das SK. Roles of Flavonoids in Plants. International Journal of Pharmaceutical Science and Technology. 2011;**6**:12-35

[33] Wang ZL, Wang S, Kuang Y, Hu Z.M, Qiao X, Ye M. A comprehensive review on phytochemistry, pharmacology, and flavonoid biosynthesis of Scutellaria baicalensis. Pharmaceutical Biology. 2018;**56**:465-484.

[34] Williams JS, Thomas M, Clarke DJ. The gene stlA encodes a phenylalanine ammonia lyase that is involved in the production of a stilbene antibiotic in Photorhabdus luminescens TT01. Microbiology. 2005;**151**:2543-2550

[35] Barros J, Dixon RA. Plant Phenylalanine/Tyrosine Ammonia-lyases. Trends in Plant Science. 2020;**25**:66-79

[36] Cheng GW, Breen PJ. Activity of Phenylalanine Ammonia-Lyase (PAL) and Concentrations of Anthocyanins and Phenolics in Developing Strawberry Fruit. Journal of the American Society for Horticultural Science. 1991;**116**:865-869

[37] Wohl J, Petersen M. Functional expression and characterization of cinnamic acid 4-hydroxylase from the hornwort Anthoceros agrestis in Physcomitrella patens. Plant Cell Reports. 2020;**39**:597-607

[38] Wang Y, Chen S, Yu O. Metabolic engineering of flavonoids in plants and microorganisms. Applied Microbiology and Biotechnology. 2011;**91**:949-956

[39] Park SR, Yoon JA, Paik JH, Jung WS, Ban Y-H, Kim EJ, et al. Engineering of plant-specific phenylpropanoids biosynthesis in Streptomyces venezuelae. Journal of Biotechnology. 2009;**141**:181-188

[40] Wang A, Zhang F, Huang L, Yin X, Li H, Wang Q, et al. New progress in biocatalysis and biotransformation of flavonoids. Journal of Medicinal Plant Research. 2010;**4**:847-856

[41] Du F, Zhang F, Chen F, Wang A, Wang Q, Yin X, et al. Advances inmicrobial heterologous production of flavonoids. African Journal of Microbiology Research. 2011;**5**:2566-2574

[42] Hwang EI, Kaneko M, Ohnishi Y, Horinouchi S. Production of plant-specific flavanones by *Escherichia coli* containing an artificial gene cluster. Applied and Environmental Microbiology. 2003;**69**:2699-2706

[43] Austin MB, Noel JP. The chalcone synthase superfamily of type III polyketide synthases. Natural Product Reports. 2003;**20**:79-110

[44] Leonard E, Koffas MAG. Engineering of artificial plant cytochrome P450 enzymes for synthesis of isoflavones by *Escherichia coli*. Applied and Environmental Microbiology. 2007;**73**: 7246-7251

[45] Leonard E, Yan Y, Fowler ZL, Li Z, Lim C-G, Lim K-H, et al. Strain improvement of recombinant *Escherichia coli* for efficient production of plant flavonoids. Molecular Pharmaceutics. 2008;**5**:257-265

[46] Santos CNS, Koffas M, Stephanopoulos G. Optimization of a heterologous pathway for the production of flavonoids from glucose. Metabolic Engineering. 2011;**13**:392-400

[47] Koopman F, Beekwilder J, Crimi B, van Houwelingen A, Hall RD, Bosch D, et al. De novo production of the flavonoid naringenin in engineered Saccharomyces cerevisiae. Microbial Cell Factories. 2012;**11**:155

[48] Middleton EJ. Effect of plant flavonoids on immune and inflammatory cell function. Adv. Exp. Med. Biol. 1998;**439**:175-182

[49] Gorai D, Jash SK, Roy R. Flavonoids from Astragalus Genus. International Journal of Pharmaceutical Sciences and Research. 2016;**7**(7):2732-2747

[50] Gangopadhyay A, Chakraborty S, Jash SK, Gorai D. Cytotoxicity of Natural Flavones and Flavonols Against Different

Cancer Cells. Journal of the Iranian Chemical Society. 2022;**19**:1547-1573

[51] Narayana KR, Reddy MS, Chaluvadi MR, Krishna DR. Bioflavonoids classification, pharmacological, biochemical effects and therapeutic potential. Indian Journal of Pharmacology. 2011;**33**:2-16

[52] Middleton E. The flavonoids. Trends in Pharmacological Sciences. 1984;**5**:335-338

[53] Yao LH, Jiang YM, Shi J, Tomás-Barberán FA, Datta N, Singanusong R, et al. Flavonoids in food and their health benefits. Plant Foods for Human Nutrition. 2004;**59**:113-122

[54] Rice-Evans CA, Miller NJ, Paganga G. Structureantioxidant activity relationships of flavonoids and phenolic acids. Free Radical Biology & Medicine. 1996;**20**:933-956

[55] Wollenweber E, Dietz VH. Occurrence and distribution of free flavonoid aglycones in plants. Phytochemistry. 1981;**20**:869-932

[56] Coultate TP. Food-The Chemistry of its Components. 4th ed. Cambridge, UK: The Royal Society of Chemistry; 2002. pp. 195-200

[57] Pierpoint WS. Flavonoids in the human diet. In: Plant Flavonoids in Biology and Medicine: Biochemical, Pharmacological and Structure-Activity Relationships. New York, USA: Alan R. Liss; 1986. pp. 125-140

[58] Swain T. The evolution of flavonoids. In: Plant Flavonoids in Biology and Medicine: Biochemical, Pharmacological, and Structure-Activity Relationships. New York, USA: Alan R. Liss; 1986. pp. 1-14

[59] Cook NC, Samman S. Flavonoids---Chemistry, metabolism, cardioprotective effects, and dietary sources. The Journal of Nutritional Biochemistry. 1996;**7**:66-76

[60] Shirley BW. Flavonoid biosynthesis: "new" functions for an "old" pathway. Trends in Plant Science. 1996;**1**:377-382

[61] Agati G, Azzarello E, Pollastri S, Tattini M. Flavonoids as antioxidants in plants: location and functional significance. Plant Science. 2012;**196**:67-76

[62] Zhao J, Dixon RA. The "ins" and "outs" of flavonoid transport. Trends in Plant Science. 2010;**15**:72-80

[63] Perez-Gregorio MR, Regueiro J, Barreiro CG, Otero RR, Gandara JS. Changes in antioxidant flavonoids during freeze-drying of red onions and subsequent storage. Food Control. 2011;**22**:1108-1113

[64] Ferdinando MD, Brunetti C, Fini A, Tattini M. Flavonoids as antioxidants in plants under abiotic stresses. In: Ahmad P, Prasad MNV, editors. Abiotic Stress Responses in Plants: Metabolism, Productivity and Sustainability. New York, USA: Springer; 2012. pp. 159-179

[65] Hatier JHB, Gould KS. Foliar anthocyanins as modulators of stress signals. Journal of Theoretical Biology. 2008;**253**:625-627

[66] Tattini M, Galardi C, Pinelli P, Massai R, Remorini D, Agati G. Differential accumulation of flavonoids and hydroxycinnamates in leaves of Ligustrum vulgare under excess light and drought stress. The New Phytologist. 2004;**163**:547-561

[67] Erlejman AG, Verstraeten SV, Fraga CG, Oteiza PI. The interaction of flavonoids with membranes: potential

determinant of flavonoid antioxidant effects. Free Radical Research. 2004;**38**:1311-1320

[68] Taylor LP, Grotewold E. Flavonoids as developmental regulators. Current Opinion in Plant Biology. 2005;**8**:317-323

[69] Jansen MAK. Ultraviolet-B radiation effects on plants: induction of morphogenic responses. Physiologia Plantarum. 2002;**116**:423-429

[70] Kuhn BH, Geisler M, Bigler L, Ringli C. Flavonols accumulate asymmetrically and affect auxin transport in Arabidopsi. Plant Physiology. 2011;**156**:585-595

[71] Mathesius U. Flavonoids induced in cells undergoing nodule N organogenesis in white clover are regulators of auxin breakdown by peroxidase. Journal of Experimental Botany. 2011;**52**:419-426

[72] Rana DK, Saha D, Sengupta PS, Sarkar B, Mondal P, Gorai D, et al. Computational and Docking Studies of 5,6-dihydroxy-7,8,4′-trimethoxyflavone: A α-Glucosidase Inhibitory Constituent from Limnophila indica. Journal of Chemistry and Chemical Sciences. 2018;**8**:548-561

[73] Naoumkina M, Dixon RA. Subcellular localization of flavonoid natural products. Plant Signaling & Behavior. 2008;**3**:573-575

[74] Panche AN, Diwan AD, Chandra SR. Flavonoids: an overview. Journal of Nutritional Science. 2016;**5**:1-15

[75] Kim JS, Kwon CS, Son KH. Inhibition of ά-glucosidase and ά-amylase by luteolin, a flavonoid. Bioscience, Biotechnology, and Biochemistry. 2000;**64**:2458-2461

[76] Oberdorster E, Clay MA, Cottam DM, Wilmot FA, McLachlan JA, Milner MJ. Common phytochemicals are ecdysteroid agonists and antagonists: A possible evolutionary link between vertebrate and invertebrate steroid hormones. The Journal of Steroid Biochemistry and Molecular Biology. 2001;**77**:229-238

[77] Salunke BK, Kotkar HM, Mendki PS, Upasani SM, Maheshwari VL. Efficacy of flavonoids in controlling *Callosobruchus chinensis* (L.) (Coleoptera: Bruchidae) a post-harvest pest of grain legumes. Crop Protection. 2005;**24**:888-893

[78] Goławska S, Sprawka I, Łukasik I, Goławski A. Are naringenin and quercetin useful chemicals in pest-management strategies? Journal of Pesticide Science. 2014;**87**(1):173-180

[79] Schnarr L, Segatto ML, Olssona O, Zuin VG, Kümmerer K. Flavonoids as biopesticides—Systematic assessment of sources, structures, activities and environmental fate. Science of The Total Environment. 2022;**824**:153781

[80] Ateyyat M. Impact of flavonoids against woolly apple aphid, eriosoma lanigerum (Hausmann) and its sole parasitoid *Aphelinus mali* (Hald.). The Journal of Agricultural Science. 2012;**4**:227-236

[81] Monique S, Simmonds J. Importance of flavonoids in insect-plant interactions: Feeding and oviposition. Phytochemistry. 2001;**56**:245-252

[82] Dorman SC, Ballard SA. Insecticidal composition containing an aromatic unsaturated carbonyl compound. 1949. Patent number US2465854A

[83] Kydonieus AF. Method and compositions for controlling flying insects. 1980. Patent number US4193984A

[84] Saotome K. Protecting method of field crop. 1982. Patent number JPS57120501A

[85] Saotome K. Crop protection method by means of Cinnamaldehyde. 1984. Patent number FR2529755A1

[86] Miyawaki H, Saotome K. Insect pest repellent. 1989. Patent number JPH01261303A

[87] Armstrong GL, Dunn-Coleman NS, Wach M. Control of fungal diseases in the production of mushrooms. 1992. Patent number US5149715A

[88] Emerson RW, Jr Crandall BG. Use of flavonoid aldehydes as insecticides. 1996. Patent number WO1996020594A1

[89] Zobitne KA, Gehret MJ. Insecticidal compositions and method of controlling insect pests using same. 2003. Patent number US6548085B1

[90] Bennett J, Brandt A, Borovsky D. Compositions and methods for controlling pests. 2003. Patent number US6593299B1

[91] Bessette SM, Beigler MA. Pesticidal activity of plant essential oils and their constituents. 2005. Patent number US6841577B2

[92] Brucker D, Sweeney M, Breen T. Methods of producing a functionalized coffee. 2005. Patent number US20050031761A1

[93] Schmidt-Dannert C, Watts K. Microorganisms for the recombinant production of resveratrol and other flavonoids. 2005, 2009. Patent number US20050208643A1, US7604968B2

[94] Nappa A, Lorenzini F, Sanhueza A. Compositions and methods to add value to plant products, increasing the commercial quality, resistance to external factors and polyphenol content thereof. 2007. Patent number US20070232495A1

[95] Koffas M, Leonard E, Yan Y, Chemler J. Production of flavonoids by recombinant microorganisms. 2008. Patent number US20060019334A1

[96] Wallace R.G. Flavonoid concentrates. 2008. Patent number US20080274519A1

[97] Dixon RA, Liu C-J, Deavours B. Methods and compositions for production of flavonoid and isoflavonoid nutraceuticals. 2010. Patent number US7750211B2

[98] Dixon RA, Modolo LV, Peel G. Production of proanthocyanidins to improve forage quality. 2011. Patent number NZ580217A

[99] Jones A. Pesticidal compositions and methods of use thereof. 2012. Patent number US8142801B2

[100] Mouradov A, Spangenberg G. Manipulation of flavonoid biosynthetic pathway. Patent number AU2010306410A1, AU2010306410B2, 2012, 2015.

[101] Yoshikazu PB, Mason TJ. Flavonoid 3′,5′ hydroxylase gene sequence and method of use thereof. 2012. Patent number JP5002848B2

[102] Aguilar MS, Romero YMH, Narvaez CMR. Pesticide made of isoquinoline alkaloids, flavonoids and vegetable and/or essential oils. 2013. Patent number MX2011010032A

[103] Mouradov A, Spangenberg G. Modification of flavonoid biosynthesis in plants. 2013, 2017. Patent number US20130340118A1, US9567600B2

[104] Ripley I. Plant protection composition and method. 2014. Patent number WO2014122446A1

[105] Bessette SM, Lindsay AD. Pesticidal compositions containing rosemary oil and wintergreen oil. 2014. Patent number US8877219B2

[106] Philippov SV, Bogorodov IM. Method and agrochemical composition for using larch wood extracts in agriculture. 2014. Patent number US 2014/0335210A1

[107] Blankenshi L, Habib A, Kang Y, Semones S. Compositions and methods for enhancing plant growth. 2015, 2016.Patent number AU2013338110A1, AU2013338110B2

[108] Gallifer N, Meeker M, Carwin S, Boltlick K, Choisy P. Anthocyanin coloured composition. 2015, 2015. Patent number CN104640461A, CN104640461B

[109] Romero YMH, Narvaez CMR, Aguilar MS, Pesticide having an insecticide, acarcide and nematicde action based on isoquinoline alkaloids and flavonoids. 2015. Patent number US2015/0216181A1

[110] Yanhua W, Liao SY, Hui JX. Application of flavonoid compound the aflavanoside II to prevention and treatment of plant nematode diseases. 2015. Patent number CN104823979A

[111] Gerardi AR, Method of extracting tobacco-derived O-methylated flavonoid and use thereof. 2015, 2020.Patent number EP2906055A1, EP2906055B1

[112] Romero H, Narvaez R, Aguilar S. Pesticide with insecticide, acaricide and nematiciated action based on isoquinolinic alcaloids and flavonoids. 2015. Patent number ES2464642B1

[113] Kurisawa M, Lee F, Chung JE, Chan PYP. Methods for producing crosslinked flavonoid hydrogels. 2016. Patent number US9439886B2

[114] Spangenberg G, Sawbridge TI, Ong E-K, Emmerling M. Manipulation of flavonoid biosynthesis in plants. 2016. Patent number US9523089B2

[115] Luo J, Butelli E, Jones J, Tomlinson L, Martin CR. Methods and compositions for modifying plant flavonoid composition and disease resistance. 2017. Patent number US9580725B2

[116] In-Gyu K, Chang-Gil K, Yeol K, Mok-Hyeong L, Hae Y, Jeong G, Choi HY, Han-Jeung C, Jae-Doo J, Hwan-Sang Y. Natural composition and method for manufacturing the composition avoiding and/or controlling the Hemiptera. 2017. Patent number KR101802249B1

[117] Lamb RD, Johnson MD. Agricultural compositions and applications utilizing essential oils. 2018. Patent number US9949490B2

[118] Tsivion Y. Pesticide containing antioxidants. 2019. Patent number WO2019016806A1

[119] Bailey TV. Arthropod pest trapping device, system and method. 2019. Patent number US10285393B2

[120] Klykov AG, Murugova GA, Timoshinova OA, Borovaya SA, Chaikina EL. Method for increasing content of flavonoids in buckwheat fruits. 2020. Patent number RU2729743C1

[121] Ishida S, Inai K, Tanaka M, Nomoto T. Plant growth regulating agent. 2021. Patent number EP3912470A1

Chapter 2

Importance of Flavonoid as Secondary Metabolites

Shuchi Dave Mehta, Sukirti Upadhyay and Gopal Rai

Abstract

Flavonoids are broad-spectrum secondary metabolites with cosmetics, pharmaceutical, nutraceutical, and medicinal applications. They play a crucial role in life span shortening complications, including diabetes, CVS disorder, and cancer. They are the secondary metabolites essential natural products due to their anti-inflammatory, antioxidant, anticancer, anti-cholinesterase, disease combating, antimicrobial, hepatoprotective, neuroprotective, cardioprotective, antiallergic, and many more pharmacological activities causing substantial economic and social burdens. They have the ability to scavenge superoxide, hydroxyl, and lipid radicals. They are a group of polyphenolic compounds having 15 carbon skeleton consisting of two benzene rings with heterocyclic pyran ring, which are classified as anthocyanins, flavonols, isoflavonols, and flavanones, and present in vegetables, fruits, flowers, seeds, stems, and leaves.

Keywords: flavonoids, classification, phytochemistry, pharmacological, cosmeceutical, nutraceutical applications

1. Introduction

Flavonoids are the largest group of naturally occurring phenols as phytochemicals. They have the capacity to occur both in the free state and as glycosides (largest naturally occurring phenols). They are the polyphenolic compounds that are biosynthesized by the polypropanoid pathway having a precursor as a phenylalanine molecule. The Latin word "flavus" means yellow, which is responsible for colors in flowers, fruits, and leaves. They are widely distributed in plants having color component properties. They are commonly distributed in the plant kingdom, bryophytes, and in pteridophytes, but not distributed in algae.

The precursor of biosynthesis of flavonoid involves condensation of 2 units of malonyl CoA, 1 unit of Acetyl CoA, and cinnamic acid (biosynthesized by shikimic acid) that results in C_{15} intermediate, which results in various kinds of flavonoids.

Flavonoids are considered as a major group of plant polyphenols having potential for cosmeceuticals and biomedical applications. They are present in food materials and plants. They are responsible for protection against pathogens, herbivores, and also ultraviolet radiation. They are absorbed from the small intestine and the colon for complete absorption. They are widely used as chemotaxonomic markers and belonging to Polygonaceae, Umbelliferae, Rutaceae, Rosaceae, Leguminosae, Lamiaceae, and also Compositae. Solubility properties include soluble in water and alcohol, whereas insoluble in organic solvents [1].

Pharmacological properties include anti-inflammatory, antiallergic effects, antithrombotic, vasoprotective, antioxidant activity, diuretic, antispasmodic, antibacterial, antifungal properties, and many more.

Quercetin, kaempferol, quercitrin, flavones, dihydroflavons, flavans, flavonols, anthocyanidins, proanthocyanidins, calchones, catechins, and leucoanthocyanidins are some of the classes that show biological and pharmacological activities, such as anticancer, antimicrobial, antistress, antiallergic, oestrogenic activity, vascular activity, and hepatoprotective activity. The antioxidant property is due to the suppression of reactive oxygen species (ROS) formation. They are also involved in the inhibition process of enzymatic activity in reactive oxygen species synthesis.

S.No.	Identification test	Procedure	Observation
1	Ammonia test	Filter paper strip was dipped in the alcoholic solution of extract. Ammoniated with ammonia solution.	Color changed from white to orange.
2	Shinoda/Pew Test	Test solution (5 ml) + 5 ml. 95% alcohol + few drops of conc. HCl + 0.5 g magnesium turning.	Appearance of pink color.

Table 1.
Identification of flavonoids.

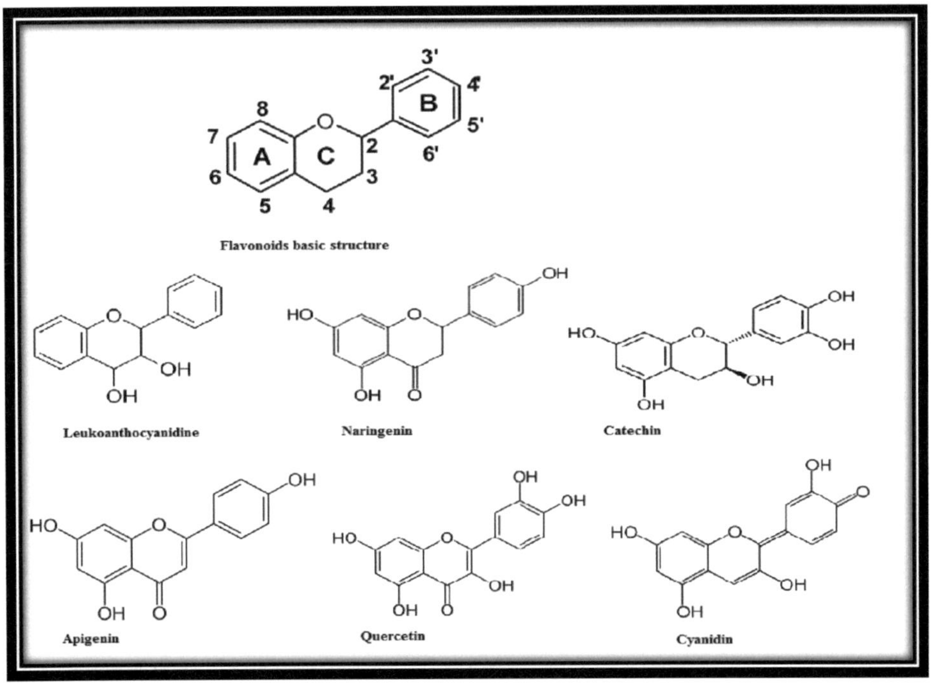

Figure 1.
Structure of basic flavonoids and their various types.

Hepatoxicity is generally treated by flavonoids, such as quercetin, catechin, rutin, and venoruton. They act by increasing hepatic Gclc expression by increasing cAMP levels, which increases Gclc and helps in the transcription. The increased Gclc expression helps in the depletion of hepatic ROS levels and proapoptotic signaling, which protects the hepatic cells. Clinical trials have shown a significant effect of flavonoids in the treatment of liver diseases. **Table 1** shows various chemical tests of flavonoids.

Naringenin, apigenin, galangin, isoflavones, chalcones, and flavanones are the flavonoids that have been observed to have antibacterial activity might be due to complex formation with protein by bonding *viz.* hydrophobic effects or covalent bond that inhibit DNA synthesis and RNA synthesis. Vegetables and fruits are the major source of flavonoids and have been observed to cure and prevent cancer. Flavonoids produce anticancer activity may be due to inhibition of tyrosine kinase activity, which is one of the growth factor signaling to the nucleus present in the cell membrane. Flavonoids inhibit fatty acid synthase activity and lipogenesis in the prostate cancer cell. It may also be due to inhibition of cell cycle arrest or inhibition of heat shock protein or inhibition of nuclear type II estrogen binding sites. Quercetin, genistein, daidzein, epigallocate-chin-3-gallate, biochanin, and hesperidin are flavonoids acting as anticancer [2].

1.1 Classification

The classification is based on the degree of oxidation of the central pyran ring where **Figure 1** shows a basic structure of flavonoids with the structure of each example of various types of flavonoids. The classification is as follows:

1. Flavandiol (3, 4-Hyroxyflavane), for example, leukoanthocyanidine.

2. Flavanones (4-Oxo-flavane), for example, naringenin.

3. Flavanols (3-Hydroxy-4-Oxo-flavane, Catechine), for example, catechin.

4. Flavones (4-oxo-flav-2-ene), for example, apigenin.

5. Flavonols (3-Hydroxy-4-Oxo-flav-2-ene), for example, quercetin.

6. Flavylium (Anthocyanidin), for example, cyanidin.

2. Flavonoids as "cosmeceutical"

Cosmetics products are products that are intended to apply on hair and skin to enhance appearance, promote attractiveness, and beautify and cleanse the properties of skin and hair. In 1990, the term "Cosmeceuticals" was described as the over-the-counter skin care products that involve therapeutic properties by addition of plant active ingredients, such as alpha-hydroxy acid, retinoic acid, ascorbic acid, and coenzyme Q10, which helps in skin elasticity, reduction of wrinkles in skin as antiaging effect, to check degradation of collagen, and also protection against UV radiation. The presence of multi-active properties in flavonoids provides protection to skin blood vessels, which telangiectasias and petechias cause by rupturing blood vessels. The main activity in cosmetic is capillary permeability reduction, blood vessel protection, and platelet aggregation prevention [3].

Phytochemicals, such as phenolic compounds, substances structurally characterized by having one or more hydroxyls attached to an aromatic ring, are classified into simple phenolics and polyphenols, which may be subdivided into tannins and flavonoids. Both flavonoids and non-flavonoids are associated with various interesting cosmetic properties, such as photoprotection, antiaging, moisturizing, antioxidant, astringent, anti-irritant, and antimicrobial activity [4].

The polyphenol nature is responsible for the colors in many fruits, vegetables, and flowers, which protect from environmental stress that act as an antioxidant. Flavonoid in cosmetics provides antioxidant protection against UV radiation protection for our skin [2, 5].

Southeastern Asian medicinal plant *Alpinia galanga* belonging to Zingiberaceae in traditional medicine used to relieve stomach pain, indigestion, and to treat skin diseases. Antioxidant, anti-inflammatory, and antibacterial are medicinal properties. Flavonoids, phenolic acids, and volatile compounds are present as phytochemicals in several parts, such as leaves, seeds, and rhizomes. Flavonoid in *A. galanga* plays a crucial role in the cosmetic area. Isolated and identified flavonoids from seeds and the rhizomes of *A. galanga* are found as 11 flavonols, 4 dihydroflavonols, one flavan 3-ol, and flavanone. Galangin (3,5,7-trihydroxyflavone) is the largest compound of the 11 flavonols, whereas kaempferol, quercetin, and myricetin are popular non-methylated flavonols. Alpinone, pinobanksin 3-acetate, and 3-cinnamate are three dihyroflavonols isolated and identified in rhizomes (**Figure 2**).

In Thailand, Vietnam, and many Southeast Asian countries, *A. galanga* was used traditionally as a major ingredient in cosmetics, which include body soap and skin care products, by developing its extract using an easy hot extraction method where water was used as a solvent. **Figure 2** shows the leaves, rhizome, and whole plant of *A. galangal*.

Recently, people use to order online the dried plant and its extract easily. In cosmeceuticals, the authentication process and identification of plants should be carried out previously for the development processes. The research on the potential of extracts and/or phytochemicals from this medicinal plant is insufficient; a greater number of studies focusing on flavonoid identification of the potential extracts from this medicinal plant should be conducted. Easy availability, easy cultivation, and low price are major reasons to promote the flavonoid bioactive ingredients of *A. galanga* in the field of cosmetics. The anti-wrinkle and antioxidant properties in cosmetics are always remembered [6].

Water lily (*Nymphaea lotus* L.) is a traditional ornamental medicinal and cosmetic plant having perennial aquatic flowering nature which is in many countries of Asia and Africa, especially in Thailand, Nepal, Vietnam, Indonesia, China, Bangladesh, and Sri Lanka. Roots, rhizome, stolon, petiole, young leaves, and flower parts were used traditionally as homemade natural cosmetics products, such as skincare and perfume, whereas medicinally in the treatment of circulatory system syndrome. It is also considered as the symbol of the Hindu Goddess "Sarasvatiji" and "Laxmiji." A high content of flavonoids in flowers help local people to use the ethanolic extract for homemade cosmetic products, especially for skincare and perfumery. Several publications have specifically revealed the potential in cosmetic and cosmeceuticals potential of *N. lotus* L. Flower and stamen are considered as the richest source of flavonoids, which conclude as important bioactive constituents for cosmetic applications. Chalcone glycoside chalcononaringenin-2"-O-galactoside, flavonol glycosides, isorhamnetin-7-O-galactoside, isorhamnetin-7-O-xyloside, isorhamnetin-3-O-xyloside, myricetin-3-O-xyloside quercetin-3-O-rhamnoside, quercetin-3-O-xyloside,

Figure 2.
Leaves, rhizome, and whole plant of Alpinia galanga.

Figure 3.
(i) Nymphaea *lotus, (ii)* Nymphaea caerulea, *(iii)* Nelumbo nucifera, *and (iv)* Nelumbo lutea.

and kaempferol-3-O-galactoside are major flavonoids present in the *N. lotus* L. For antioxidant activity, *in vitro* assays include DPPH (2,2-diphenyl-1-picrylhydrazyl), ABTS {2,20-azino-bis(3 ethylbenzothiazoline-6-sulfonic acid)}, FRAP (Ferric Reducing Antioxidant Power Assay), BHT (Butylated hydroxytoluene), cellular assays and animal studies were performed and showed a positive result [7].

Anti-wrinkle, skin whitiliser, coolant, anti-acne, and relaxing activities are reported. The anti-acne property is due to antibacterial properties. The literature revealed that *N. lotus* L. has properties to improve skin complexion, reduce skin pigmentation, and improve skin with soothing and emollient activity.

The genera lotus of varieties *N. lotus*, *Nymphaea caerulea, Nelumbo nucifera,* and *Nelumbo lutea* have potential that can be utilized in herbal cosmetics due to the presence of a large amount of flavonoids, which reported as anti-acne, skin whitiliser, and anti-pimple activity with improve skin texture. **Figure 3** shows various varieties of *N. lotus*.

The market formulation includes Veet hair removal gel, AHAVA mineral botanic body lotion, Clarisonic daily acne cleanser, Lotus sunscreen, Neutrogena deep cleanser, face wash, skin tonner, and many more, which showed a positive response from people and increase the demand in the cosmetic industry. Mechanism of action includes inhibition of tyosinase which results in skin whitelising, inhibition of elastase and DOPA oxidase inhibition which results in antiwrinkle activity, antiradical property prevents inhibition of Ultraviolet radition which prevent skin tanning, and inhibition of melanin which also results skin whitilising and anti-ageing effects [8].

3. Flavonoids as nutraceuticals

Nutraceuticals may range from isolated nutrients, dietary supplements, diets to genetically engineered “designer” food, herbal products, and processed products, such as cereals, soups, and beverages. A nutraceutical is any nontoxic food extract supplement that has scientifically proven health benefits for both the treatment and prevention of diseases. According to Stephen DeFolice, nutraceuticals are food or parts of food that provide medical or health benefits, including the prevention and treatment of disease. In simple language “nutraceuticals” are food materials utilized for treatment and prevention of disease and may range from isolated nutrients, dietary supplements, diets to genetically engineered “designer” food, herbal products, and processed products, such as cereals, soups, and beverages. Flavonoid rich foods are considered as superfoods nutraceuticals include plant origin food mainly tea, fruits, grains, legumes, nuts, vegetables, and wine.

Flavonoids rich foods as nutraceuticals have explored the working mechanisms which include pharmacological activities, such as anti-wrinkle, antiaging, anticancer, antibacterial, hypoglycemic, anti-hypertension, anti-obesity, antiproliferative, anti-thrombotic, and anti-platelet aggregation. The presence of phenolic compounds is confirmed by potent antioxidant activity and metal chelators agent. Daily diets consumed are always rich in these flavonoids. Major active ingredients that are considered in the plant are flavonoids that have a long half-life with less side effects and are absorbed in the intestine after ingestion. They have a high absorption capacity in the intestine.

The mechanisms behind the activities are trapping of free radicals, decreasing leukocyte immobilization, and regulation of nitric oxide and xanthine oxidase activity. The pharmacokinetic studies of flavonoids are shown in **Figure 4**, which includes absorption, distribution, and biotransformation where hydrolysis is an important part of absorption in the cecum and colon by enterobacteria, aglycone is absorbed by gut epithelial cells and enter the circulation to metabolize in the liver [8–11].

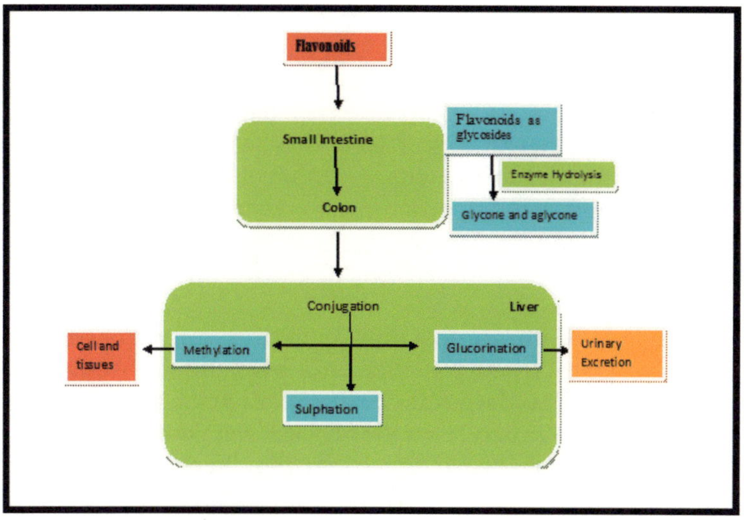

Figure 4.
Pharmacokinetic studies of flavonoids.

4. Flavonoids as pharmaceuticals

Flavonoids are the bioactive phytochemical constituents that are present in fruits, herbs, stems, cereals, nuts, vegetable, flower, and seeds and give biological activities. About 10,000 flavonoid compounds are isolated and identified, showing effective antioxidant, anticancer, antibacterial, cardioprotective agents, anti-inflammation, immune system promoting, and skin protectant for medical application.

4.1 Anticancer activity

Cancer is a major health problem that can be defined as impaired cell cycle and uncontrolled proliferation, which results in the growth of abnormal cells. Increased exposure to stress, pollution, radiation, ultraviolet rays, smoking, oxidative stress, genetic mutation, and lack of apoptotic function are the major causes of cancer. Anthocyanins, flavones, flavones, flavonols, and chalcones are major flavonoids having anticancer activities. The mixed mechanism of action of flavonoids as anticancer action includes down-regulation of mutant p53 protein, inhibition of expression of Ras proteins, estrogen receptor binding capacity, tyrosine kinase inhibition and cell cycle arrest. Quercetin was reported to exert a growth inhibitory effect on various tumor cells also cell cycle arrest in proliferating lymphoid cells. *Dryopteris erythrosora Erythrina suberosa, Phaseolus vulgaris* L., *Medicago truncatula* Gaertn, *Ceratonia siliqua* L., *Butea monosperma, Glycyrrhiza glabra* L. and many more plants reported as anticancer due to presence of flavonoids [12–14].

4.2 Antioxidant activity

Plants, animals, and human protection against the effect of reacting oxygen species by suppressing reacting oxygen species with chelation of the trace elements involved in the free radical generation and enzyme inhibition. They are found in the chloroplast having scavenging activity of singlet oxygen and stabilizers of the chloroplast outer envelope membrane [15].

The enzymatic and non-enzymatic systems involve detoxification and removal of oxidant species of glutathione [GSH], GSH peroxidase, GSH reductase, and GSH S-transferase. The prooxidant activity of flavonoids becomes cytotoxic that undergoes transition metal reactions resulting in the formation of highly reactive oxygen species that damages protein and DNA [1].

4.3 Antimicrobial activity

According to World Health Organization, multidrug-resistant pathogenic microorganisms are major global health complications, which involve various natural products. As one of the class of secondary metabolites of the natural class, flavonoids play a vast and crucial role to handle multidrug-resistant pathogenic microorganisms strains with their versatile pharmacological activities. Prenylation or geranylation at C6; and hydroxylation of C5, C7, C3', and C4' have reported to enhance bacterial inhibition of flavonoids, whereas methoxylation at C3' and C5 has been studied to decrease antibacterial action of flavonoid. It is reported that the cell membrane is found at the major site of flavonoid action, which help in the inhibition of the respiratory chain and the ATP synthesis that also involves damage to phospholipid bilayers. Flavanone is acting as potent antibacterial activity by synthesizing a compound with halogenations of the B ring, as well as lavandulyl or geranyl substitution of the A ring [16–18].

4.4 Cardioprotective activity

More than 4000 flavonoids, which include chalcones, flavonols, dihyroflavonols, catechins, isoflavones, and catechins have the capacity for cardioprotective activity against myocardial ischemia or reperfusion as antihypertensive, anti-atherosclerotic, and anti-platelet. The significant role of flavonoids by preventing cardiovascular diseases, which may be due to antioxidant, antithrombotic, and antiatherogenic activity. For example, red wine consumption will protect against thrombosis and atherosclerosis by inhibition of platelet aggregation and LDL oxidation. The literature revealed that a daily diet of 100 mg of flavonoid helps in the reduction and possibility of cardiovascular diseases by inhibition of low-density lipoprotein oxidation and reduced platelet aggregability [19–22].

5. Conclusion

For thousands of years, flavonoids in plants are utilized as traditional medicine. They are polyphenolic compounds that are biosynthesized by the polypropanoid pathway and have potential as cosmeceutical, nutraceutical, and pharmaceutical applications. They work in the multi-mechanism of action, such as protecting endothelial cell, inhibiting foam cell formation, regulating lipid metabolism, anti-inflammatory, and underlying molecular mechanism. In future prospective, flavonoids as nutraceutical, cosmeceutical, and pharmaceutical aspects would help to reduce the burden in urban and rural populations of developed and developing countries by elevation of its pharmacokinetic, metabolic, and pharmacodynamic characteristics and using in novel drug delivery system technology.

Author details

Shuchi Dave Mehta[1*], Sukirti Upadhyay[2] and Gopal Rai[1]

1 Guru Ramdas Khalsa Institute of Science and Technology-Pharmacy, Jabalpur, India

2 School of Pharmaceutical Sciences, IFTM University, Moradabad, India

*Address all correspondence to: shuchi1801@gmail.com

References

[1] Costa R, Costa Lima SA, Gameiro P, Reis S. On the development of a cutaneous flavonoid delivery system: Advances and limitations. Antioxidants (Basel, Switzerland). 2021;**10**(9):1376. DOI: 10.3390/antiox10091376

[2] Arct J, Pytkowska K. Flavonoids as components of biologically active cosmeceuticals. Clinical Dermatology. 2008;**26**(4):347-357. DOI: 10.1016/j.clindermatol.2008.01.004

[3] Sumit K, Vivek S, Sujata S, Ashish B. Herbal cosmetics: Used for skin and hair. Inventi Rapid: Cosmeceuticals. 2012;**2012**(4):1-7

[4] Oliveira CA, Dario MF. Bioactive Cosmetics. In: Martínez L, Kharissova O, Kharisov B, editors. Handbook of Ecomaterials. Springer, Cham; 2018. DOI: 10.1007/978-3-319-48281-1_157-1. ISBN: 978-3-319-48281-1

[5] Epstein H. Cosmeceuticals and polyphenols. Clinical Dermatology. 2009;**27**(5):475-478. DOI: 10.1016/j.clindermatol.2009.05.011

[6] Tungmunnithum D, Tanaka N, Uehara A, Iwashina T. Flavonoids profile, taxonomic data, history of cosmetic uses, anti-oxidant and anti-aging potential of Alpinia galanga (L.) willd. Cosmetics. 2020;7:89. DOI: 10.3390/cosmetics7040089 1-8

[7] Tungmunnithum D, Kongsawadworakul P, Hano C. A cosmetic perspective on the antioxidant flavonoids from Nymphaea lotus L. Cosmetics. 2021;**8**:12. DOI: 10.3390/cosmetics8010012; 1-9

[8] Sable NV, Pagar SA. A review on Lotus: Use in herbal cosmetics. Research Journal of Topical and Cosmetic Sciences. 2013;**4**(1):81-83

[9] Waheed Janabi AH, Kamboh AA, Saeed M. Flavonoid-rich foods (FRF): A promising nutraceutical approach against lifespan-shortening diseases. Iranian Journal of Basic Medical Sciences. 2020;**23**(2):140-153. DOI: 10.22038/IJBMS.2019.35125.8353

[10] Jain PK, Kharya MD, Gajbhiye A, Sara UVS, Sharma VK. Flavonoids as nutraceuticals. A review. Tropical Journal of Pharmaceutical Research. 2008;7(3):1089-1099

[11] Tapas AR, Sakarkar DM, Kakde RB. Flavonoids as nutraceuticals: A review. Tropical Journal of Pharmaceutical Research. 2008;7(3):1089-1099

[12] Kumar S, Pandey AK. Chemistry and biological activities of flavonoids: An overview. The Scientific World Journal. 2013;**2013**:16 Article ID: 162750. DOI: 10.1155/2013/162750

[13] Kopustinskiene DM, Jakstas V, Savickas A, Bernatoniene J. Flavonoids as anticancer agents. Nutrients. 2020;**12**(2):457. Published 2020 Feb 12. DOI: 10.3390/nu12020457

[14] Greenwell M, Rahman PK. Medicinal plants: Their use in anticancer treatment. International Journal of Pharmaceutical Sciences and Research. 2015;**6**(10):4103-4112. DOI: 10.13040/IJPSR.0975-8232.6(10).4103-12

[15] Roy A, Khan A, Ahmad I, Alghamdi S, Rajab BS, Babalghith AO, et al. Flavonoids a bioactive compound from medicinal plants and its therapeutic applications. BioMed Research

International. 2022;**9**:5445291. DOI: 10.1155/2022/5445291

[16] Shamsudin NF, Ahmed QU, Mahmood S, Ali Shah SA, Khatib A, Mukhtar S, et al. Antibacterial effects of flavonoids and their structure-activity relationship study: A comparative interpretation. Molecules. 2022;**27**:1149. DOI: 10.3390/molecules27041149

[17] Yuan G, Guan Y, Yi H, et al. Antibacterial activity and mechanism of plant flavonoids to gram-positive bacteria predicted from their lipophilicities. Scientific Reports. 2021;**11**:10471. DOI: 10.1038/s41598-021-90035-7

[18] Cushnie TP, Lamb AJ. Antimicrobial activity of flavonoids. International Journal of Antimicrobial Agents. 2005;**26**(5):343-356. [published correction appears in Int J Antimicrob Agents. 2006;27(2):181. DOI: 10.1016/j.ijantimicag.2005.09.002

[19] Khan J, Deb PK, Priya S, Medina KD, Devi R, Walode SG, et al. Dietary flavonoids: Cardioprotective potential with antioxidant effects and their pharmacokinetic, toxicological and therapeutic concerns. Molecules. 2021;**26**:4021

[20] Testai L, Martelli A, Cristofaro M, Breschi MC, Calderone V. Cardioprotective effects of different flavonoids against myocardial ischaemia/reperfusion injury in Langendorff-perfused rat hearts. The Journal of Pharmacy and Pharmacology. 2013;**65**(5):750-756. DOI: 10.1111/jphp.12032

[21] Cook NC, Samman S. Flavonoids—Chemistry, metabolism, cardioprotective effects, and dietary sources. The Journal of Nutritional Biochemistry; 1996;7(2):66-76. ISSN: 0955-2863, DOI: 10.1016/S0955-2863(95)00168-9

[22] Kumar MD, Gustav J, Thomas R. The significance of Ayurvedic medicinal plant. Journal of Evidence-based Complementary and Alternative Medicine. 2017;**22**(3):494-501

Chapter 3

Flavonoids Biosynthesis in Plants as a Defense Mechanism: Role and Function Concerning Pharmacodynamics and Pharmacokinetic Properties

Asmaa Nabil-Adam, Mohamed E. Elnosary, Mohamed L. Ashour, Nehad M. Abd El-Moneam and Mohamed A. Shreadah

Abstract

Flavonoids are a major class of secondary metabolites that comprises more than 6000 compounds that have been identified. They are biosynthesized via the phenylpropanoid metabolic pathway that involves groups of enzymes such as isomerases, hydroxylases, and reductases that greatly affect the determination of the flavonoid skeleton. For example, transferase enzymes responsible for the modification of sugar result in changes in the physiological activity of the flavonoids and changes in their physical properties, such as solubility, reactivity, and interaction with cellular target molecules, which affect their pharmacodynamics and pharmacokinetic properties. In addition, flavonoids have diverse biological activities such as antioxidants, anticancer, and antiviral in managing Alzheimer's disease. However, most marine flavonoids are still incompletely discovered because marine flavonoid biosynthesis is produced and possesses unique substitutions that are not commonly found in terrestrial bioactive compounds. The current chapter will illustrate the importance of flavonoids' role in metabolism and the main difference between marine and terrestrial flavonoids.

Keywords: marine flavonoids, biosynthesis, pharmacodynamics, pharmacokinetics, defense mechanism

1. Introduction

Flavonoids with more than 6000 individuals are divided into six main categories: chalcones, flavones, flavonols, flavandiols, anthocyanins, and proanthocyanidins are present in all plants. The aurones group is present in several species [1]. Legumes and a few non-legume plants produce isoflavonoids, but few plants produce

3-deoxyanthocyanins and phlobaphenes. Stabileneds resemble chalcones and are made from grape and peanuts [2]. Flavonoids have several functions, such as protecting plants from UV radiation and phytopathogens, regulating signals, promoting male fertility, transporting auxin, and giving flowers their color to draw pollinators [3]. Flavonoids may increase nutrient recovery during senescence by shielding leaf cells from photooxidative damage. The oldest and most prevalent flavonoids are flavonols, which have potent physiological effects [4]. The phenylpropanoid pathway, which turns phenylalanine into 4-coumaroyl-CoA, produces flavonoids. Chalcone scaffolds are produced by the first flavonoid-specific enzyme, chalcone synthase. Although the principal method for producing flavonoids in plants is consistent, the different flavonoid subclasses are produced depending on the species via isomerases, reductases, hydroxylases, and various Fe2+/2-oxoglutarate-dependent dioxygenases [4]. Transferases alter the solubility, reactivity, and interaction of flavonoid molecules with biological targets by adding sugars, methyl groups, and acyl moieties to the flavonoid backbone [5]. Plants can produce specific organic molecules and prevent metabolic interference thanks to metabolic channeling. P450s-related metabolons have been discovered in several biosynthetic pathways, including phenylpropanoid, flavonoid, cyanogenic glucoside, and others [6]. More proof of intermediate channeling is provided by transgenic tobacco plants that produce two phenylalanine ammonia-lyase isoforms (PAL1 and PAL2) and cinnamate-4-hydroxylase [7]. For example, Yeast-two hybrid assays indicate that rice contains an anthocyanin multienzyme complex [8]. For flavonoids from marine environments, for example, unknown and uncommon marine flavonoids precursors of algal flavonoids are biosynthesized using comparable metabolic mechanisms to those seen in plants. Flavonoids are created by several metabolic pathways [9]. There are a variety of structures in algal flavonoids. Specify flavones, isoflavones, flavanols, flavanones, and favonols. The C_6/ C_3 unit of t-cinnamic acid and the unit of malonyl-C_3 CoA make up the backbone. P-coumaric acid is present in *Anabaena doliolum, Spongiochloris spongiosa, Porphyra tenera,* and *Undaria pinnatifida* [10]. Phenylalanine Ammonia Lyase is responsible for producing t-cinnamic acid. While p-coumaroyl-CoA may be converted into a chalcone derivative by Claisen condensation, Michael addition, and chalcone synthase-catalyzed enolization, the poly-b-keto ester can be made using the phenylpropanoid pathway. Chalcone is created from the poly-b-keto ester [11] Chalcone is transformed into flavonoid structures via several reductases, isomerases, hydroxylases, acyltransferases, and glycosyltransferases. Within each category, structural heterogeneity is brought on by variations in the number of linked hydroxyl groups, position, degree, type of alkylation, and glycosylation [12].

Further evidence is required for the algal flavonoid synthesis route. For Algal flavonoid composition, the 15-carbon skeleton of flavonoids comprises two phenyl rings (A and B) connected by a 3-carbon unit to form a heterocyclic ring. Their structural categorization is based on where the benzenoid substituent is located [13]: 2-Phenylchromans (flavonoids), which include anthocyanidins, flavanones, flavonols, flavones, and flavan-3-ols. 2. Pterocarpans, isoflavones, and 3-phenylchroman isoflavones. Glycosides represent the majority of flavonoids in marine [14]. Phlorotannins are more often used polyphenols in algae than flavonoids. There is not much literature on these chemicals. Flavanols. The most varied flavonoids found in algae are flavanols. The double bond between carbons 2 and 3 and the carbonyl group in carbon 4 of ring C is absent from flavanols. Occupational and health advantages Polyphenols have a variety of bioactivities in addition to being antioxidants that protect against UV rays and are poisonous to predators [15]. The bioactivity of phlorotannins and flavonoids is regulated by the pattern of -OH group substitutions, double bonds, and the site of their conjugation [16]. Also, Phytoflavonoids bioactivity of the flavonoids found in algae is unclear, but there are several applications for marine flavonoids in medical and cosmetic applications. Acanthophorin A and B, isolated from *A. spicifera*, shield the rat liver from lipid peroxidation and stop the malondialdehyde generation [17]. Because OH groups combine with H radicals to generate persistent semiquinone radicals, flavonoids have antioxidant properties. Flavonoids scavenge hydroxyl, other functional groups, and unsaturated and conjugated pi bonds [18]. Flavonoids are vital components of the human diet and potent antioxidants that can lower oxidative stress and several diseases in people [19]. Due to their antioxidant, anti-inflammatory, antibacterial, and affinity/inhibitory properties toward inflammatory enzymes, plant extract rich in flavonoids are employed in dermatology and cosmetics. As dietary components, algal flavonoids may protect against several human diseases (**Figure 1** and **Table 1**) [25].

2. Flavonoids as a defense mechanism

Flavonoids are necessary for plant development and plaque resistance. Flavonoids are responsible for many of the hues of angiosperm flowers. They can be found throughout the plant, not just in the blooms [26]. Plant-based foods and drinks like fruits, vegetables, tea, chocolate, and wine are rich in flavonoids. Plants, animals, and even microorganisms contain flavonoids. Flavonoids responsible for The color and aroma of flowers, fruit dispersal, the germination of seeds and spores, and the growth and development of seedlings in plants are all influenced by flavonoids, which are produced in particular locations [27]. In addition to acting as UV filters [28], signal molecules, allopathic chemicals, phytoalexins, detoxifying agents, and antimicrobials, flavonoids shield plants against biotic and abiotic stresses. Plants' ability to adapt to heat and tolerate freezing may be influenced by flavonoids [29]. Early advances in

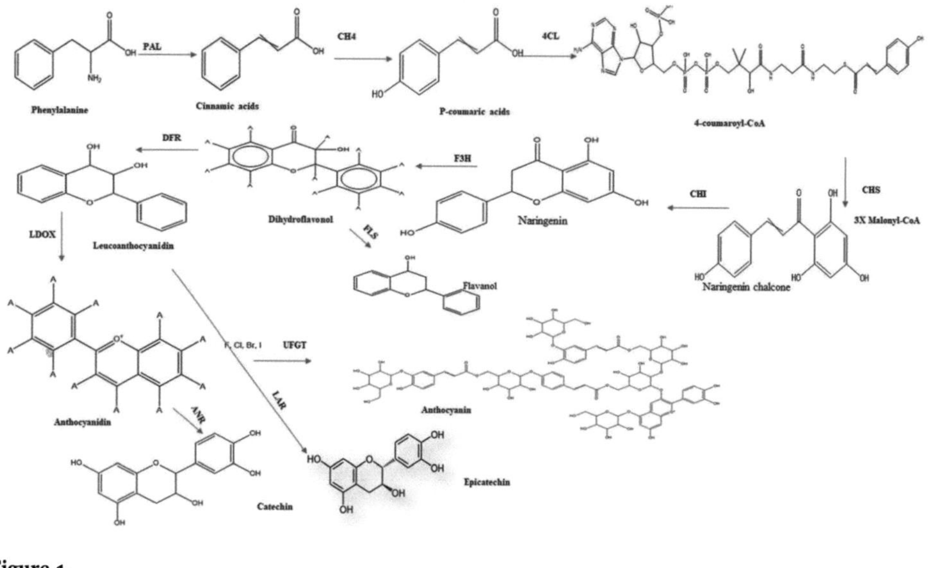

Figure 1.
The metabolic pathways of flavonoids.

Flavonoids classes	Basic structure	Description	Activity	Examples of subclasses	References
Anthocyanins	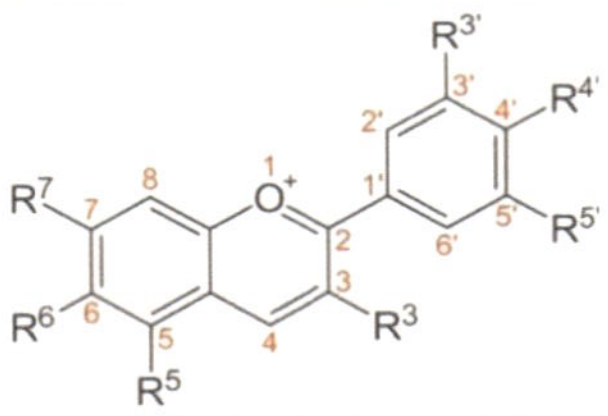	Anthocyanin is derived from flavanol, and it has the basic structure of flavylium ion, that is, a lack of ketone oxygen at the 4-position	Anthocyanins are polyphenols with known antioxidant activity, which may be responsible for some biological activities, including preventing or lowering the risk of cardiovascular disease, diabetes, arthritis, and cancer.	• cyanidin, • delphinidin, • pelargonidin, • peonidin, • petunidin, • malvidin	[20]
Chalcones	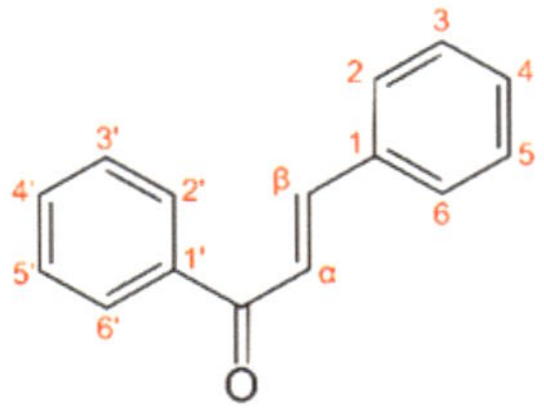	Chalcones are α,β-unsaturated ketones (trans-1,3-diaryl-2-proper-1-ones), consisting of two aromatic rings (A and B) attached by an α,β-unsaturated carbonyl system with a variety of substituents	Chalcones and their analogs possess significant biological activities, including antimicrobial, anticancer, antitubercular, antioxidant, anti-inflammatory, antileishmanial, enzyme inhibitory, and miscellaneous applications, and hence acquire a unique place in medicinal chemistry.	• butein • cardamonin • sappanchalcone • 3-deoxysappanchalcone • isobavachalcone • bavachalcone • ajejuchalcone • morachalcone A • broussochalcone A • xanthohumol • xanthoangelol • gemichalcone C	[21]
Flavonols	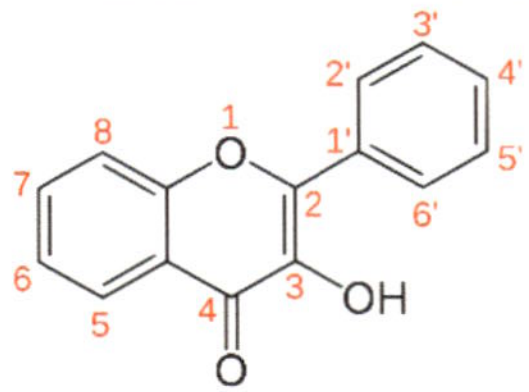	Flavonols are the most abundant yellow colorants. There are a variety of different sources of flavonols, but only a few of them produce a sufficient quantity of dyes with relatively good color fastness	Flavonols are abundant in many plant species and, like flavonols, have been shown to have antioxidant and antimicrobial activities	• kaempferol, • quercetin, • myricetin • fisetin • Aglycones • Galangin • Morin • Myricetin • Rhamnetin • Glycosides • Quercitrin • Rutin	[1]

Flavonoids classes	Basic structure	Description	Activity	Examples of subclasses	References
flavanones	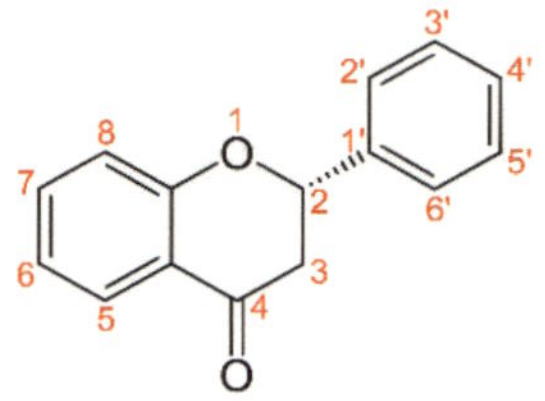	The flavanones, a type of flavonoids, are various aromatic, colorless ketones derived from flavones that often occur in plants as glycosides.	Anticancer, antioxidants. Antimicrobial, anti-inflammatory	• Blumeatin • Butin • Eriodictyol • Hesperetin • Hesperidin • Homoeriodictyol • Isosakuranetin • Naringenin • Naringin • Pinocembrin • Poncirin • Sakuranetin • Sakuranin • Sterubin • Pinostrobin	[22]
Flavones	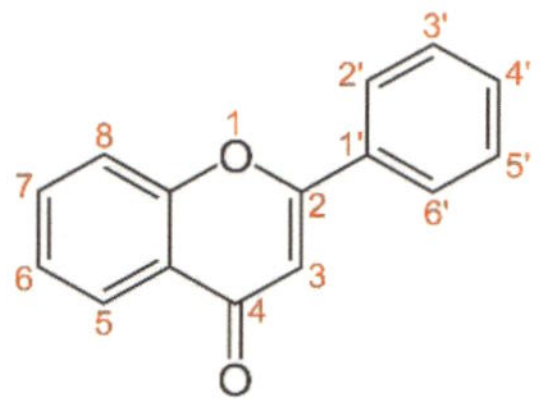	Flavones are common in foods, mainly from spices and some yellow or orange fruits and vegetables.	Among them, natural flavones, as well as some of their synthetic derivatives, have been shown to exhibit several biological activities, including antioxidant, anti-inflammatory, antitumor, antiallergic, neuroprotective, cardioprotective, and antimicrobial.	• Primuletin • Chrysin • Tectochrysin • Primetin • Apigenin • Acacetin • Genkwanin • Echioidinin • Baicalein • Oroxylin A	[23]
Isoflavonoids	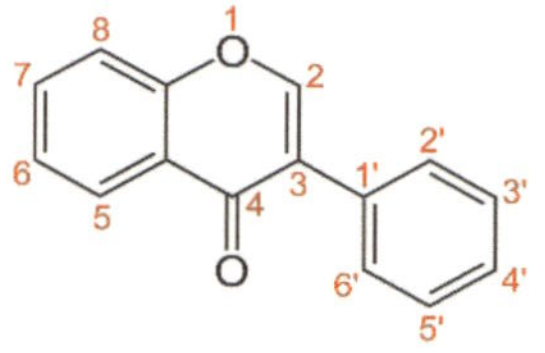	Isoflavonoids have been classically defined as dietary antioxidants, i.e., compounds that may protect against oxidative stress linked to inflammation and the risk of macromolecule damage by free radicals and by related oxygen and nitrogen-based oxidizing agents	The potential health benefits of isoflavones may include protection against age-related diseases, including cardiovascular disease, osteoporosis, hormone-dependent cancer, and loss of cognitive function. The mechanisms involved may include weak oestrogenic action and antioxidant activity.	• genistein, • daidzein • , glycitein	[24]

Table 1.
The different classes of flavonoids with their activities and examples of each class.

floral genetics were made possible by mutation techniques that altered flower colors generated from flavonoids, and functional gene silencing in plants was associated with flavonoid synthesis. Today, flavonoids treat illnesses and prevent cancer, and more than 6000 flavonoids color fruits, herbs, vegetables, and medicinal plants. Flavones are a subclass of flavonoids. Positions 2 and 3 of the C ring have double bonds, and position 4 is a ketone [30], Most flavones found in fruits and vegetables have a hydroxyl group at position 5 of the A ring, but other positions—particularly position 7 of the A ring and 3′ and 4′ of the B ring—can vary depending on the taxonomic group. Keto-flavonoids are flavanols. Flavanols can act as antioxidants and reduce the risk of vascular disease [31]. The third hydroxyl group on the C ring of flavonols can be glycosylated. Compared to flavones, flavonols exhibit distinct methylation, hydroxylation, and glycosylation patterns. The most varied group of bioactive polyphenols is flavonoids [32]. A phenyl ring (A ring) joined with heterocyclic benzo-c-pyrone (C ring), which connects to another phenyl ring (B ring) via a carbon-carbon bond, makes up the three rings that make up the diphenyl propane skeleton of flavonoids ($C_6C_3C_6$). These chemicals contain hydroxyls. The gymnosperms, angiosperms, ferns, and bryophytes contain more than 4000 flavonoids [2]. The first plant to possess flavonoids is green algae [33]. According to Bonfante [34], a symbiotic relationship between algae and a tip-growing fungus is the reason for the plants' biphyletic origin. When plants transitioned from marine to terrestrial habitats, flavonoids developed primarily to protect against rising UV exposure [35].

Research on plant flavonoid production is an important area of research. The synthesis of flavonoids in algae may differ from higher plants due to algae development. Microalgae contained flavonoids and flavonoid intermediates by Goiris et al. [36]. Phloretin and dihydrochalcone might be intermediates in the production of flavonoids. The findings suggest that flavonoid biosynthesis enzymes may be present in microalgae. Diverse flavonoids compatible with better plant flavonoid synthesis are present in certain algae [36]. Plant-plant interactions may be impacted by flavonoids. Negative relations are mainly based on the inhibition of seedling development and germination. Flavonoids are frequently released into the soil by roots, where they prevent seed germination. They may also be found in leaves and pollen, which prevents the germination of other plants [3, 37]. Barley flavones lessen weed seed germination, while *Centaurea maculosa* catechins limit *Centaurea diffusa* and *Arabidopsis thaliana* germination and growth [38]. The precise allelopathic mechanism of flavonoids is unknown. Allelopathy can be affected by preventing cell division, ATP production, and auxin activity [39]. The Ca^{2+} signal cascade and root system death are stimulated by flavanols. Due to its ability to inhibit weed development, allelopathy is becoming increasingly important in agriculture [40].

Plants may fight against bacteria and fungi with the assistance of flavonoids. The general antipathogenic properties of flavonoids are largely attributed to their antioxidant properties. They suppress ROS produced by both pathogens and plants [41]. The B ring of flavonoids can intercalate or form hydrogen bonds with nucleic acid bases, limiting bacterial DNA and RNA synthesis and influencing DNA gyrase activity [42]. They can bind to viral nucleic acids or capsid proteins and inhibit viral polymerases [43]. The antipathogenic activity of flavonoids depends on their structure. Flavones and flavanones without substitution have strong antifungal properties. The antifungal activities of these compounds are reduced by hydroxyl and methyl groups, but methylated flavonoids have a greater effect. Isoflavones, flavanes, and flavanones are powerful antibacterial compounds, whereas flavonoids inhibit root infections, particularly fungus (**Figure 2**) [44, 45].

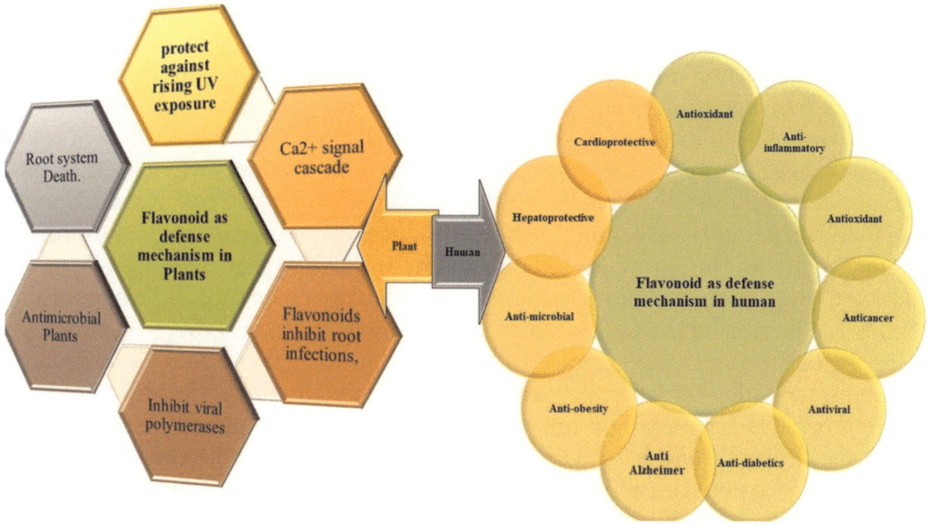

Figure 2.
The different role of flavonoids and their mechanism of action as a defense mechanism in human Vis plant.

3. Role and function of flavonoids as a protective preventive and curative effect against various diseases

Therapeutic flavonoids are associated negatively with sickness, according to epidemiological research. Conventional flavonoids can interact with key enzyme systems and show polypharmacological action. It follows that the considerable study of chemical structure-activity connections is not surprising. Strong antiviral properties of bioactive flavonoids, including those against the hepatitis C virus, and antimicrobial such as *Escherichia coli*, have been examined. Chemical processes, including methoxylation, glycosylation, and hydroxylation, have been mostly responsible for these effects. Research on the structure-activity relationship (SAR) covers several elements. C_2C_3 double bonds are frequently advantageous—hydroxylation substitution style is important [46–49].

A beneficial role for 5−/7-hydroxyl derivatives in ring A hydroxylation is suggested by six anti-H5N1 influenzas A virus 5, 7-diOH flavonoid candidates, and daidzein's less potent anti-human fibroblast collagenase catalytic domain (MMP1ca) activities. Better ring B hydroxylation indicates stronger MMP1ca inhibition by 3′-OH and 5′-OH drugs. Catechol is the most common functional group. Innovative drug production has been stimulated by quercetin, more notable than morin inhibition of canine distemper virus [50, 51]. Compared to luteolin, quercetin considerably contributes to ring C. It also affects how many hydroxyl groups there are. More hydroxyl groups lessen the hydrophobicity of flavonoids, preventing membrane partitioning. Hydrophobicity and electronic delocalization impact the intensity of hydroxylation, which causes some hydroxyl-rich flavonoids to act more strongly. Different hydroxyl groups may raise C_3 charges while decreasing hydrophobicity, which suggests pharmacological activity. Methylation hurts membrane fluidity and lowers the activity of several viruses and bacteria according to their physiology. Two PMFs performed less well against *E. coli* than equivalent aglycone. Antiviral activities can be found in flavonoid glycosides [52–54]. Finding the right screening substances for dietary therapy and medical treatment may result from analyzing the SAR behaviors displayed by certain flavonoids in antiviral/bacterial situations. Apoptosis induction, proteasome inhibition, nuclear factor signaling suppression, differentiation induction, cell cycle arrest induction, receptor contact, and interaction with carcinogenic enzymes are a few of the mechanisms that have highlighted the importance of flavonoids in cancer therapy. Flavonoids have potential as anticancer medications since they can selectively kill cancer cells. The molecular planarity and conjugation between rings C and A/B that the C_2C_3 double bond produces are necessary to prevent tumor growth. Studies on the C_2C_3 double bond and its anticancer properties have been conducted using tumor cell lines, such as colon adenocarcinoma cells. Stronger inhibition was obtained by the C_2C_3 unsaturation and two hydroxyl groups on ring B [55, 56].

Numerous studies have demonstrated how hydroxylation affects tumor regulation. Per-methoxylated flavonoids do not have the same anticancer effects as hydroxylated flavonoids. To, 6-OH and 5, 7-diOH contribute, Ring B does not become less active when hydroxyl groups are added [1]. Ring C's 3-hydroxylation enhances its biological actions. Without 3-OH, flavonoids have less antiproliferative activity. The 3-OH molecule's affinity for the binding site might be greater [30, 57]. Methylated flavonoids support the enhanced biological action of ringing A polymethoxylation. Among the Organ flavonoids tested in the cell morphology research, two A-ring PMFs exhibit the highest proliferative inhibition, demonstrating the significance of the C-8 position in flavonoids' antiproliferative impact. The bioactivity of flavonoids against neurodegeneration has traditionally been linked to their antioxidant properties. However, new research has highlighted the significance of acetylcholinesterase (AChE) and butyrylcholinesterase (BChE) interactions, mitochondrial dysfunction, key neuronal signaling pathways, and chelation of transition metals in controlling neuronal resistance to neurotoxic oxidants and inflammatory mediators. Ring B hydroxylation may improve learning to avoid cardiovascular disease.

Effects of flavonoids and the eNOS transcription factor the second Krüppel-like component [58, 59]. The C_2C_3 double bond produces an effective twofold structure in eNOS and ET-1 synthesis, and 4-carbonyl moiety results in about 1.35-fold higher gene expression (quercetin vs. epicatechin/catechin), according to the findings. The "protein-binding" mechanism was highlighted in a SAR examination of 12 flavonoids with paraoxonase1 (rePON1) due, at least in part, to different hydroxylation substitutions, the C2C3 double bond, and the 4-carbonyl group in ring C. Flavones and flavonols have stronger PON1 interactions because of the C_2C_3 double bond in ring C, which increases molecular planarity and may cause electron delocalization between rings A and B. Coplanarity exists between the 3-hydroxyl group and the 4-carbonyl oxygen atom. Flavonoids' greatest therapeutic benefits are in managing leukemia, sepsis, asthma, and other inflammatory diseases. SAR research is more important because flavonoids have been extensively investigated and certain mechanisms may not be unified. Double bonds in C_2C_3 might encourage molecular planarity. For example, Hesperetin's absence results in a lower volume/surface ratio than diosmetin's absence [19, 50, 60]. (b) Isoflavones, the ring B catechol moiety, are subject to 3′- and 5′-hydroxylations that promote cell differentiation. (c) Ionizing hydroxyl groups and blocking the NF-kB signaling pathway during methylation boost the anti-inflammatory impact. (d) Because of their decreased hydrophobicity and sterical hindrance, glycosides with lower lipophilicity have fewer anti-inflammatory effects [50, 61]. Moreover, a large replacement has been researched. Anti-inflammatory flavonoids have three taxonomic markers: the C-butyrolactone moiety, the 5-acetic acid/lactone group, and the C_7C_8 double bond. DM is a sophisticated hyperglycemic condition. The flavonoids that fight diabetes are widely recognized. By hydroxylation and planarity at position 7, several flavonoids can activate PPAR. Methoxylation enhances the antidiabetic efficacy of flavonoids on 3 T3-L1 adipogenesis, but hydroxylation has a detrimental effect [62–64]. SAR is the substitution of glycosylation, particularly glucosylation at position 3. C-3-Glu/detail. More proof is required on Gly's mechanism of glucose regulation. Transition metals that promote radical hydroxyl formation in reduced forms through the Fenton reaction can be bound by flavonoids. Due to the resorcinol moiety of ring A, isoflavone has the highest antioxidant activity among studied flavonoids. These numbers indicate the structural elements of antioxidants. SAR is aided by the C_2C_3 double bond conjugated to a 4-carbonyl group in the flavonoid subclasses ring C. According to some authors, there is no clear connection between these moieties and antioxidant function when other structural conditions are satisfied. Despite cellular ROS inhibition and structural moieties being similar, certain flavonols have strong electron-donating action [32, 65, 66]. A C_2C_3 double bond conjugated to a 4-carbonyl group improves antioxidant activity when other structural requirements are satisfied. The dissociation constant of phenolic hydroxyl groups and the stability of phenoxy radicals in ring B are both impacted by the 4-carbonyl group's propensity to create electron shifts through resonance effects. Ring C and A/B can be conjugated thanks to the electron coupling and molecular planarity provided by the unsaturated C_2C_3 double bond. Likewise, 5-OH creates hydrogen bonds. 4-Carbonyl delocalizes the ring B electron, increasing the antioxidant effect when combined with the C_2C_3 double bond or other electron-donating groups. The degree and location of hydroxylation affect how anti-oxidative flavonoids are. Stable flavonoid radicals are created when hydroxyl groups on the ring B absorb hydrogen and electrons. Two hydroxyl groups in ring B considerably increase antioxidant activity [67–70]. The primarily responsible pharmacophore is the 3′, 4′-catechol group, which generates an ortho-semiquinone radical by electron delocalization and confers high activity through intra-molecular hydrogen bonding between catechol hydroxyl groups. Outside of the two hydroxyl groups on ring B, no one substitution makes sense. With 4′-OH, apigenin promotes erythroid differentiation. Higher inhibitory effects were seen from flavonoids with an ortho-dihydroxyl group in ring B than those with a 4′-hydroxylation. In comparison to ring A's meta-dihydroxylation, ring B's ortho-dihydroxyl group is more easily oxidized 5, 7-di-OH in ring A inhibits the activity of antioxidants. Strong 5- and 7-OH as 2, 4-substituted resorcinol substructure activities are highlighted by luteolin, quercetin, kaempferol, and apigenin [71–73]. There is proof that alterations in ring A's positions 5 and 7 that donate electrons prevent the 3-hydroxylation of ring C. 3-OH inhibits antioxidant activity compared to luteolin and quercetin. When examining overall hydroxylation, both the electron transit within the resonance system and the total hydroxyl groups are considered. The flavonoid nucleus with more hydroxyl groups is held up in the hydrophobic cavity because hydrophilicity rises with the number of hydroxyl groups [18, 74, 75]. The antioxidant activity is decreased by altering the methylation of the free hydroxyl groups on ring B. Methoxyl flavonoid derivatives have higher antioxidant activity due to flavonoid-flavonoid interaction [76, 77]. Ring A's several methoxylation substitutions should offset the catechol moiety of ring B. Antioxidant activity of flavonoid O- or C-glycosides has been investigated. Chemical tests have shown that c-glycosides are more effective antioxidants than O-glycosides. Compared to O-glycosides, C-glycosyl flavonoids exhibit about 100% higher radical scavenging action [53, 73, 78]. Another experiment shows that the antioxidant activity of c-glycosides is roughly 50%. The aforementioned C-glycoside studies still require in vivo information and in-depth analysis. Flavonoid glycosides develop in food as A- or C-ringed O-glycosides. The author speculates that the sugar moiety in position 3 could exacerbate steric hindrance or polarity. In addition, ring A's antioxidant capabilities are enhanced by 6-glucosylation but diminished by 8-glucosylation [54, 79].

4. Pharmacokinetics and pharmacodynamics of flavonoids

4.1 Pharmacokinetic characteristics of flavonoids

There are several pharmacological functions for flavonoids. However, problems impede their approval as prescription drugs for usage in clinical settings and, to some extent, future research. Plant yield, bioavailability, and low solubility are problems. For 20 years, researchers have studied the metabolism and absorption of flavonoids. The distribution, metabolism, excretion, toxicity, and absorption of flavonoids are not optimal and differ between classes [80]. Flavonoids' in vivo concentration is decreased by their low oral absorption. Low solubility, little oral absorption, and significant phase-I and phase-II hepatic enzyme metabolism. Chemical interactions between bacteria and small intestine epithelial cells affect flavonoid absorption due to intestinal metabolism. In small-intestine epithelial cells, flavonoids are glucuronidated, O-methylated, and sulfated, which reduces their bioactivity [81–84]. In rats, only 20% of oral quercetin was absorbed; the rest was converted to CO2 and excreted in the feces. Within 48 hours, the body excretes absorbed quercetin [85]. The PK profile of flavonoids in plants is influenced by light, temperature, oxygen exposure, pH, and UV radiation. The synthesis of plant flavonoids can be altered by UV light. Flavonoids' extraction and shelf-life are influenced by temperature. 45 to 60°C are ideal for extracting flavonoids from the tissue of the pericarp of litchi fruit [86, 87].

It is challenging to link a single flavonoid molecule to pharmacological action since flavonoids are available as a plant extract that includes several plant natural components. The PK profile of certain flavonoid changes, such as methylation and glycosylation, can be improved. In the next section, we shall talk about how methylation and glycosylation impact the pharmacokinetics and bioavailability of flavonoids. Derivative of Flavonoids with Better PK Properties [88, 89]. Their chemical structure governs the bioavailability and chemical stability of flavonoids [90–92].

Absorption, distribution, and metabolism are all impacted by changing the flavonoid skeleton. Methylation, which is the process of adding a methyl group to a substrate, controls cellular energy, epigenetics, and gene expression [93, 94]. Methylated flavonoids, which get a methyl group through the hydroxyl group, and C-methylated flavonoids, in which the methyl group is directly attached to the C atoms of the basic skeleton, are two different types of methylation flavonoids, depending on the location. OMT and CMT catalyze the methylation of O and C, respectively (CMT). SAM provides the methyl group through a biomolecular nuclear substitution (SN2) procedure. The first SAM-dependent methyltransferase to crystallize is catechol-OMT [95, 96]. Methylated flavonoids outperform their non-methylated analogs in terms of stability, potency, and bioavailability. Chemical characteristics, immunogenicity, and PK/PD are all influenced by glycosylation. To increase solubility, stability, and toxicity, flavonoids can be glycosylated to produce O- or C-linked glycosides. Glycosylated flavonoids can either be O-glycosides or C-glycosides, depending on the glycosidic connection to the basic flavonoid skeleton. While the sugar molecule in C-glycosides is connected to the basic flavonoid skeleton by its respective carbon atoms (generally at C-6 and C-8 positions), the sugar moiety in O-glycosides is coupled to the basic skeleton by a hydroxyl bond (commonly at 3-C and 7-C positions) [54, 89, 97]. Glycosylation often occurs in the subclasses of flavones and flavanols. Rutin and hesperidin are two flavonoids that do not dissolve well in water or alcohol. In non-polar solvents, non-glycosylated flavonoids (glycans) dissolve. Glycosylation increases the chemical stability of flavonoids in vitro. The stability of glycosylated flavonoids is increased, making them promising. A few glycosylated flavonoids, including luteolin-4′-O-glucoside and apigenin-7-glucoside, also inhibit BCRP [1, 73, 98]. Glucosylated were mostly used as examples to keep the effects of glycosylation easy to understand.

4.2 Pharmacodynamics of the polyphenolics

The major concern of several studies was the regulatory effects of polyphenolics on the human body to understand the Pharmacodynamics mechanism [99–105]. The current study investigates the effect of 14 polyphenolic compounds from the sponge, and the pharmacodynamics of the 14 compounds were listed and investigated according to ADME (Adsorption, Distribution, Metabolism, and Excretion). Not all bioavailable compounds are physiologically active. Furthermore, polyphenolic compounds' pharmacodynamics are not linked to their physiological activity [84]. For that, many studies were established to discover how common polyphenolics' pharmacodynamics, especially flavonoids, correlate with their inhibitory activities. The polyphenolic bioavailability profiles are classified into subgroups; for example, isoflavones consider the most absorbed type of flavonoids, followed by quercetin. The previous study indicates that the amounts of these compounds in plasma (after intestinal and hepatic metabolism) are very low, indicating their most flavonoids were eliminated rapidly [106]. Quercetin administered significantly inhibited platelet function and signaling in vivo studies. Physiological effects of flavonoids correlate with structural features of these compounds; there is also evidence to show that flavonoid dynamics in vivo are likely to be complex. Also, flavonoids can reduce the pathological effects of atherosclerosis, thrombosis, and CVD risk, while flavonoid bioavailability has been researched extensively [107, 108]. Quercetin is one of the most important plant molecules that has shown many pharmacological activities, such as being anticancer, antiviral, and treating allergic, metabolic, and inflammatory disorders, eye and cardiovascular diseases, and arthritis [109]. It has also shown a wide range of anticancer properties, and several reports indicate its efficacy as a cancer-preventing agent. Quercetin also has psychostimulant properties and has been documented to prevent platelet aggregation, capillary permeability, and lipid peroxidation and enhance mitochondrial biogenesis. Gallic acids mainly involved MAPK and NF-κB signaling pathways. It thus greatly reduced the inflammatory response by decreasing the release of inflammatory cytokines, chemokines, adhesion molecules, and cell infiltration [110, 111]. Thus the main Pharmacological activities and pharmacodynamics mechanism of Gallic acids were associated with anti-inflammatory effect. Pyrogallol is mostly used in pharmaceutical companies for medicinal purposes as a topical antipsoriatic. Pyrogallol showed both prooxidant and antioxidant activities. Additionally, Pyrogallol act as an antimicrobial activity by generating reactive oxygen species and is critical for its [112, 113]. Kaempferol has been confirmed to hurt cancerous cells of different types by triggering apoptosis, cell cycle arrest at the G2/M phase, downregulation of signaling pathways, and phosphoinositide 3-kinase (PI3K)/protein kinase B (AKT). Kaempferol also induces the activation of cysteine proteases involved in apoptosis initiation, preventing the accumulation of reactive oxygen species (ROS) in cancer development [114]. coumarin has previously reported a wide range of pharmacological activities, such as anticancer, anti-inflammatory, antioxidants, anti-coagulant, and antibacterial [115]. Phenolic acids have two types: hydroxybenzoic acids, such as gallic, and hydroxycinnamic acids, such as ferulic, caffeic, and o-coumaric acid. o-Coumaric acid is a hydroxycinnamic acid with different biological activities, such as anti-lipidemic, antioxidant, and anti-carcinogenic [116]. Furthermore, the therapeutic effect of o-Coumaric acid in a human breast cancer cell line (MCF7) treatment through CYP isozymes mRNA levels was reported by Sen et al. [117], that studied the effect of o-Coumaric acid on drug-metabolizing CYP enzymes at the mRNA and protein expression levels was investigated in a human hepatocarcinoma cell line (HepG2 cells and that also confirmed in the current study as the extract exhibit cytotoxicity against HepG2 with 41.2 ug. Hydroxycinnamic acids (HCAs) (coumaric acid, ferulic acid, caffeic acid, and Chlorogenic acid) in general and Chlorogenic acid acids specifically are of high importance due to their beneficial pharmacological effects [118]. HCAs are mainly recognized as potent antioxidants and have a diverse therapeutic effect against various diseases, for example, cardiovascular and neurodegenerative diseases and cancer Anti-inflammatory and antimicrobial activities. Ferulic acid inhibited the synthesis of TNF-alpha and decreased, Ferulate, their antioxidant mechanism of action through maintaining redox regulation, suppressing NF-κB activation, and modulating the expression of NF-κB-induced, pro-inflammatory such as COX-2 and iNOS [50]. Also, the NF-kB suppression by Ferulate is mediated via suppressing the activation of NIK/IKK and MAPKs. Caffeic acid inhibits the activities of COX-1 and COX-2 enzymes and inhibits prostaglandin synthesis and COX. Caffeic acid also decreased several inflammatory cytokines such as interleukin (IL)-beta, IL-6, and tumor necrosis factor (TNF)-. Chlorogenic acid has an antidiabetic and anti-obesity role and significantly decreases the level of cholesterol and triacylglycerol [119]. The same effect was observed with ferulic acid as the mechanism of action of ferulic involved the suppression and/or down-regulation of lipid metabolism genes [120]. Additionally, chlorogenic acid significantly elevated beta-oxidation and lipase activity in diabetic animals. The catechol has an anti-inflammatory role through inhabiting the NF- κB, and TNF- α [50].

5. Role of flavonoids against different diseases

5.1 Role of flavonoids as antioxidants

Phenylalanine, tyrosine, and malonate are used by plants to produce flavonoids. The flavan nucleus consists of three rings (C_6-C_3-C_6) with 15 carbon atoms each, referred to as A, B, and C. While individual compounds within a class fluctuate in A- and B-ring substitution, classes of flavonoids vary in the degree of oxidation and pattern of C-ring substitution. Interesting flavonoids include flavones, flavanones, isoflavones, flavonols, flavanonols, flavan-3-ols, and anthocyanidins. Flavonoids include biflavones, chalcones, aurones, and coumarins. Hydrolyzable tannins, proanthocyanidins (oligomers of flavan-3-ol), caffeates, and lignans are all examples of plant phenols. In vitro antioxidant activity. Gutteridge and Halliwell Antioxidants may upregulate or maintain antioxidant defenses, scavenge ROS, and reduce ROS formation by inhibiting enzymes or chelating trace elements involved in free radical creation [121–128]. The aforementioned procedures are part of flavonoid activity. Some of their effects might result from interactions between enzymes and radical scavenging. The ROS-producing enzymes microsomal monooxygenase, glutathione S-transferase, mitochondrial succinoxidase, and NADH oxidase are all inhibited by flavonoids [129]. The majority of flavonoids are potent antioxidants, which may account for their health advantages. Flavonoids (Fl-OH) can reduce highly oxidizing free radicals with redox potentials of 2.13–1.0 V due to their low redox potentials (0.23 E7 0.75 V) [130]. To do this, they donate a hydrogen atom to the reaction Fl-OH + R• Fl-O• + RH (1), where R• stands for superoxide anion, peroxyl, alkoxyl, and hydroxyl radicals. When combined with another radical, the peroxyl radical (Fl-O•) may create a stable quinone. Free radicals are scavenged by flavonoid antioxidants [131]. Strong free radical scavengers called flavonoids have drawn interest as potential therapies for diseases caused by free radicals and oxidative stress [42]. By forming complexes with them, flavonoids stabilize oxidative free radicals. Studies on the structure of flavonoids and their capacity to absorb free radicals are comprehensive. According to Kumar and Pandey [42], the heterocyclic and B ring structure and substituents affected the radical-scavenging activity. Radical-scavenging ability is determined by the presence of a catechol group in ring B, which has enhanced electron-donating properties and serves as a radical target, and a 2,3-double bond conjugated with the 4-oxo group. The heterocyclic ring's 3-hydroxyl group aids in radical scavenging, whereas the hydroxyl or methoxyl groups at positions 3,5, and 7 of rings A and C seem to be less important [132]. These structural characteristics increase the antioxidant power of flavonoids or the stability of the peroxyl radical. Both flavonols and flavones containing a catechol group in ring B are very active; flavonols, however, are more potent owing to the presence of the 3-hydroxyl group. Rutin's capacity to scavenge free radicals is decreased by glycosylation. A hydroxyl group in ring B of myricetin increases its antioxidant capacity (pyrogallol). Ring B's lone hydroxyl lowers activity [133]. Due to their weak antioxidant properties and 2,3-double bond with the 4-oxo group, flavanonols and flavanones are. The antioxidant effects of flavan monomers and flavanonols are comparable (catechin vs. taxifolin). The antioxidant potential is increased by the allocation of the 3-hydroxyl group or the incorporation of a pyrogallol group in ring B (as in epigallocatechin). If ring B includes catechol, then anthocyanidins and their glycosides (anthocyanins) are comparable to quercetin and catechin gallates (like in cyanidin). Kaempferol's antioxidant activity is decreased when the 3-hydroxyl group from ring B in pelargonidin is removed (which differs from quercetin because it has a lone hydroxyl group in ring B) [18].

5.2 Role of flavonoids as neuroprotective against different neurodegeneration diseases

There is a long history of using flavonoids in medicine. They are a notable therapeutic class because of their diversity, dispersion, and seclusion. Flavonoids are essential for the development of new drugs since they may be used as natural products and are the basis for many treatments [19]. Medicinal plants, vegetables, fruits, and wines all contain flavonoids. Flavonoids can bind to body proteins, and alter hormones, enzymes, transporters, and DNA. They can also chelate heavy metals and scavenge free radicals [134]. Numerous pharmacological studies demonstrate their efficacy in treating microbiological infections, cancer, cardiovascular diseases, neurological disorders, and diabetic Mellitus (DM) [135]. A recent study suggests that consuming foods high in flavonoids may improve cognitive abilities in humans [136].

In both normal and transgenic preclinical animal models of Alzheimer's disease (AD), certain flavonoids have been demonstrated to reduce the progression of the disease's pathology [137]. Foods high in flavonoids, such as chocolate, green tea, and blueberries, have good health effects as a result of their interactions with certain cellular and molecular targets [31]. The expression of neuromodulatory and neuroprotective proteins as well as the quantity and quality of neurons are increased by the interaction of flavonoids with ERK and PI3-kinase/Akt receptors [138]. They may improve cognitive performance by boosting blood flow to the brain and brain neurogenesis thanks to their favorable effects on the cerebrovascular system. Recently, many additional advantages of flavonoids were discovered [139]. Flavonoids lessen symptoms similar to AD and related neurodegenerative diseases [140]. Inhibiting the major enzymes involved in the development of amyloid plaques, oxidative stress, and neuronal death brought on by neuro-inflammation are a few potential treatments [141]. By preserving neuronal number and quality in key areas of the brain, flavonoids guard against diseases that impair cognitive function.

5.3 Effectiveness of flavonoids as therapeutic approach in dementia and Alzheimer's

In animal models, flavonoids reduce AD and cognitive dysfunctions, demonstrating their therapeutic utility in neurology. By focusing on important enzymes, flavonoids prevent the growth and accumulation of amyloid plaques (A). In an AD mouse model, anthocyanin-rich flavonoids in bilberry and black currant extracts lessen behavioral deficits and alter APP processing [138]. In a transgenic PSAPP animal model of cerebral amyloidosis, chronic therapy with tannic acid may enhance memory and behavior. Nobiletin improves A-mediated memory deficits and reduces A load in the hippocampi of transgenic rats. Grape polyphenols improve memory and lower soluble A oligomers in the brain tissues of Tg2576 rats. Citrus flavonoid luteolin decreases BACE1 activity and A peptide synthesis in APP transgenic neurons [139, 140]. Grape seed extracts rich in polyphenols and curcumin lessen the deposition of A in the brains of AD animals. Through the estrogen receptor, phosphoinositide 3-kinase, and Ak, Epigallocatechin gallate (EGCG) promotes non-amyloidogenic APP processing. Selective estrogen receptor modulators may be a treatment option since post-menopausal estrogen depletion is linked to an increased risk of AD. An alternative to estrogen-based therapy may be EGCG-mediated estrogen receptor modulation [141, 142]. For neuroprotective benefits, EGCG suppresses fibrillogenesis and A-rich amyloid fibrils. Unfolded polypeptides are prevented from converting directly into neurotoxic intermediates by contact with them. Big A fibrils may be split up into smaller proteins by EGCG, avoiding aggregation and negative effects. Cognitive deficits linked to neurodegeneration may benefit from myricetin's in vitro anti-amyloid activity [143–145]. These findings imply that flavonoids may inhibit the A-forming enzyme BACE1 and hence prevent the fibrillization process that leads to the generation of A. The neuro-modulating capacity and therapeutic potential of flavonoids need more investigation. With 15 carbon atoms organized into three rings, two of which are aromatic and connected by an oxygenated heterocyclic ring, flavonoids are polyphenolic chemicals that are obtained from plants. They are consumed by people in fruits, nuts, seeds, flowers, tea leaves, herbs, spices, and red wine [30, 146]. Recent research links dietary flavonoids to reduced risk of dementia, relief from neuronal degenerative conditions, and improved memory and learning. Studies reveal that these compounds' permeability across the blood-brain barrier (BBB) is

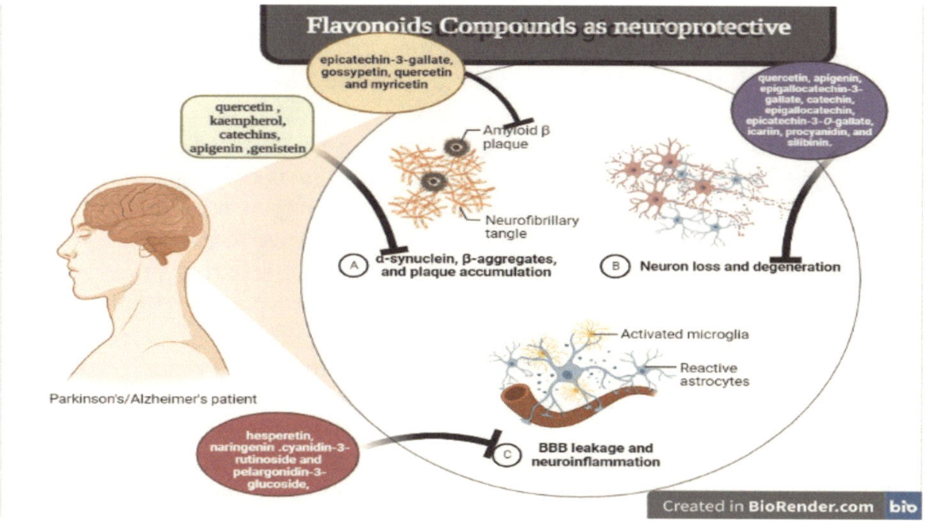

Figure 3.
Shows the protective effect of flavonoids against neurodegenerative disorders. Created with BioRender.

influenced by their structural configuration, which promotes the investigation of their therapeutic potential. Several flavonoids, including naringenin, quercetin, hispidulin, hesperetin, naringenin, and EGCG, may cross the blood-brain barrier due to their lipophilicity or interactions with BBB efflux transporters, including the P-glycoprotein. Plasma and blood flavonoids provide evidence that they may enter the brain. Following the consumption of meals or beverages high in flavonoids, human plasma contains flavonoids (**Figure 3**) [147–149].

5.4 The protective and therapeutic role of flavonoids against diabetes

Anti-diabetic flavonoids include quercetin, naringin, hesperidin, epigallocachetin gallate, apigenin, myricetin, and anthocyanins. They are antioxidants and anti-inflammatory. Flavonoids have impacts on gene regulation. Cells treated with flavonoids reveal an obscure in vivo mechanism. By controlling the activity of the intestinal carbohydrate, flavanols improve glucose homeostasis [114, 150]. Numerous studies demonstrate that the anti-apoptotic properties of flavanols increase cellular replication, insulin secretion, and glucose synthesis. As a result, catechin-rich flavanol increased insulin release prompted by glucose. Increased flavonoid consumption has been linked in human studies to a decreased incidence of diabetes. In human clinical research, flavonoids seem to not affect diabetes Consumption of isoflavones was not linked to modifications in fasting insulin, glucose, or HbA1c. Individual isoflavones seldom ever have an impact on insulin sensitivity and glycemic control [151–153]. By influencing cell mass and function, Insulin sensitivity, and glucose absorption, anthocyanins may improve glucose homeostasis. Flavonoids and Type 2 Diabetes Flavonoids are plentiful, structurally unique chemicals. Over the last ten years, the antioxidant properties of flavonoids have aided diabetic patients in reducing oxidative stress. It has been established that diabetes is caused by oxidative stress. Supplementing with antioxidants has been utilized to lessen oxidative stress caused by diabetes. Flavonoids modulate transcription factors and proinflammatory mediators and have strong in-vitro and in-vivo antioxidant and anti-inflammatory actions. T2DM may be treated by pancreatic islet isolation and transplantation [154–156]. Pancreatic transplantation and flavonoids may provide new therapeutic insights. A surplus of flavonoid molecules is also necessary. Since most flavonoids impact how complicated carbs are digested and how quickly glucose is absorbed, appropriate doses of pure single flavonoids may improve glycemia. Flavonoids fight against diabetes. As previously mentioned, signaling pathways are potential targets for treatment because they play significant roles in the pathophysiology of oxidative stress-induced diabetes. By limiting the release of cytochrome-c from mitochondria into the cytosol and inhibiting caspase activity, EGCG exhibits anti-inflammatory actions in pancreatic cells. A good target for diabetes treatment is AMPK [69, 157, 158]. As a result of EGCG's stimulation of the AMPK system, hepatic gluconeogenesis is decreased, fatty acid oxidation is improved, and mitochondrial biogenesis is controlled. In skeletal muscles, AMPK activation causes an increase in GLUT4, which facilitates glucose absorption. Hesperidin and naringin increase WAT GLUT4 while inhibiting liver GLUT2. The IRS-1-PI3-K-PKB/Akt insulin pathway was controlled by flavonoids from Oxytropis falcata Bunge chloroform extract, which decreased inflammatory cytokines by downregulating NF-B expression and increased GLUT4 expression. The flavonoid fisetin is found in foods including strawberries, apples, grapes, cucumbers, and others [159–162]. Feinstein treatment decreased glycemia, HbA1c, NF-B p65 unit, interleukin-1 beta (IL-1), and serum nitric oxide (NO) due to improved plasma insulin antioxidant status, according to animal studies. By modifying NF-epigenetics, fisetin reduced HG-triggered cytokine levels in monocytes. A diabetic dietary supplement is B's Fisetin. A flavonoid called morin may be found in wine, fruits, *Prunus dulcis*, and *Psidium guajava* [163–165]. Morin, which has anti-inflammatory properties and is useful in treating inflammatory illnesses, was demonstrated by Heeba et al. [166] to lower the cytokines IL-1, IL-6, and TNF in diabetic mice when administered at a dose of 30 mg/kg body weight. In rat liver and BRL3A cells, morin reduces fructose-induced alterations in hepatic SphK1/S1P signals and hepatic NF-B activation with IL-1b, IL-6, and TNF. The root and fruit of Scutellaria baicalensis Georgi contain a flavonoid called baicalein, which has potent antioxidant properties [166, 167]. Baicalein reduced food intake, body weight, and HbA1c levels in diabetic rats. Baicalein reduced iNOS and TGF-1 expression, inhibited NF-B, and enhanced renal tissue structure. AGEs, TNF, NF-B activation, and histopathological changes are all decreased by baicalein. Adipocytes, skeletal muscle, cardiomyocytes, and other organs all have GLUT-4. It is a glucose transporter that is resistant to insulin. Hormone/metabolic activity and tissue-specific response [168–170]. Insulin and muscle contraction cause it to move from its natural location in the cytoplasm to the plasma membrane, where it absorbs glucose. Insulin-resistant cells across the plasma membrane have changed intracellular GLUT-4. T2DM is a result of increased insulin resistance, inadequate insulin synthesis, and insulin resistance. According to a study, flavonoids and polyphenols increase GLUT-4 expression and glucose absorption. In adipocytes and skeletal muscle cells, quercetin and procyanidins enhance GLUT-4 mRNA. In mouse embryonic fibroblasts, flavonoids increase the expression of GLUT-4 mRNA. In skeletal muscle cells, epigallocatechin gallate (EGCG) elevates GLUT-4. Adipocyte and skeletal muscle cells treated with hesperidin and naringin had similar outcomes [171–173]. A flavonoid called Enicostema little increases the expression of the IRS-1, Akt-2, and GLUT-4 genes, which in turn stimulates the IRS-1/PI3K/Akt pathway. Clonal INS-1E cells and pancreatic human islets have shown protective benefits in response to kaempferol flavonoids. In human -cells and islets, 10 M kaempferol enhances viable cell concentration and inhibits apoptosis. By decreasing caspase-3 proteins and glucotoxicity and lipotoxic effects by reducing Akt and Bcl-2 anti-apoptotic activities, activating signaling pathways guards against apoptosis in cells. In chronic hyperglycemia or hyperlipidemia, −cell survival is improved by cAMP-mediated signaling [174, 175]. As previously noted, kaempferol is a special anti-diabetic chemical. Numerous studies have shown the relationship between flavonoids and glucose status. An effective treatment target for T2DM and insulin resistance is provided by flavonoids from banana flowers. Thus, IR/HepG2 cells consume more glucose when exposed to 10 mg/ml of enicostema littoral flavonoid. Hepatic insulin resistance may result from endoplasmic reticulum stress. In those with T2D, ER stress either increases insulin resistance or reduces insulin secretion [176–178]. Unfolded protein response (UPR) signaling is activated in diabetes by ER stress, which also increases inositol-requiring enzyme 1 (IRE1). Then, by converting dormant XPP-1 s into the active form, XBP1 activates IRE1. The IRE1-XBP1 pathway leads to insulin resistance by phosphorylating IRS-1. This method lowers insulin release from pancreatic islet cells. Insulin secretion is suppressed by pro-insulin mRNA degradation caused by overexpression of IRE1 [179–181]. The medicinal plant pomegranate has anti-diabetic effects. PGF has been used to treat diabetes for many years. Flavonoids, in particular, are anti-inflammatory and antioxidant polyphenols found in PGF extract. It reduces blood glucose, triglycerides, and insulin resistance, according to animal studies. By activating PPAR- 28, PGF has shown antihyperglycemic effects [182, 183].

Recently, Chinese researchers administered polyphenol extract PGF to diabetic rat models for 4 weeks at doses of 50 and 100 mg/kg. The results showed that increasing insulin-stimulated phosphorylation of IRS-1, Akt, and GSK-3 improved insulin sensitivity. The ER stress signals IRE1 and XBP-1 splicing are reduced by PGF. IRE1, XBPs, and CHOP were all decreased by PGF. PGF increases insulin resistance, which lowers glucose levels in T2DM rats, and this effect is likely mediated through Akt-GSK3 signaling and a decrease in ER stress. Unfolded protein response, mitochondrial oxidative stress, endoplasmic reticulum, and other signaling pathways all contribute to the development of T2DM. Flavonoids function as enzyme inhibitors, peroxide decomposers, hydrogen and electron donors, quenchers of singlet oxygen, radical scavengers, and metal chelators [184, 185]. These compounds regulate antioxidant enzymes, gene expressions, and protein expressions in oxidative stress-induced in vivo and in vitro models. Concentration, polarity, media, and other antioxidants all affect effectiveness. Diabetes has been associated with ER stress. Interest has grown in small molecules that inhibit ER stress and target UPR proteins [135, 186–188]. The bioavailability and bioefficacy of flavonoids may be increased by employing nanotechnologies, such as nanoparticulate systems. Flavonoids should not affect physiological processes that involve ROS. Significant antioxidant activity, ROS stability, receptor affinity, low toxicity, and free radical scavenging action are all desirable traits in flavonoids. It is important to think about the detection of ROS/RNS and the linked species and physiological levels (**Figure 4**).

5.5 The antiviral role of flavonoids against various types of virus

Viruses have envelopes made of protein, RNA, and DNA. The metabolism and environment of the host are necessary for reproduction and survival. They take advantage of host cells to spread [189]. Flavonoids are phytochemicals that inhibit viruses in a variety of ways. They could prevent DNA replication, protein translation,

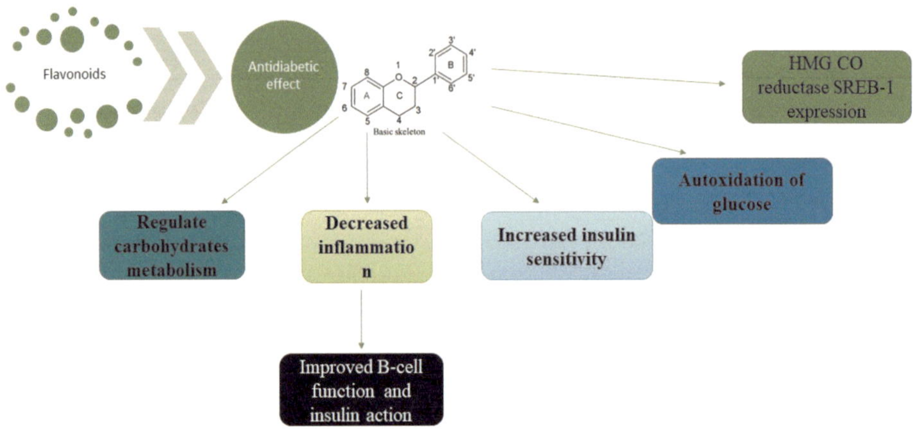

Figure 4.
Showed the protective effect of flavonoids against diabetics disorders.

processing of polypeptides, and viral attachment and entrance into cells. The invasion of healthy cells by viruses may be stopped by them. Flavonoids might bind to viral surface proteins and stop the virus from entering host cells. While other flavonoids obstruct viral assembly, packaging, and release, some of them hinder viral transcription and replication. Flavonoids modify the immune system and reduce viral load [190]. The backbone of flavones is 2-phenyl-1-benzopyran-4-one. Flavones include apigenin, baicalein, chrysin, luteolin, tangeritin, wogonin, and 6-hydroxy flavone. Since the 1990s, when it was shown that apigenin and acyclovir boosted antiviral activity against HSV-1 and HSV-2 in cell culture, flavones have been known to have antiviral properties [191]. Apigenin is effective against HSV-1, poliovirus type 2, HCV adenoviruses and hepatitis [192]. African swine fever virus (ASFV) production is decreased by 3 log due to apigenin's suppression of viral protein synthesis [193, 194]. According to Shibata et al. [195], apigenin prevents HCV replication by lowering microRNA122 which is unique to the liver. The production of early and late HCMV proteins, as well as DNA, was reduced by baicalein, but not polymerase activity. Baicalein and its analogs may be utilized to treat Tamiflu-resistant viruses, according to novel baicalein analogs with bromine-substituted B-rings that have shown substantial activity against the H1N1 Tamiflu-resistant virus [196, 197]. In early-stage infected cells, baicalin decreased HIV-1 in vitro replication. The antiviral function of baicalin prevents the HIV-1 envelope protein from interacting with immune cells [192, 196, 197]. Against the dengue virus, baicalein and baicalin (DENV). By eliminating extracellular viral particles, they prevented the growth of DENV-2.Baicalein demonstrated a strong affinity for the DENV NS3/NS2B protein (−7.5 kcal/mol) and the NS5 protein (−8.6 kcal/mol), according to in silico studies [198–201]. Baicalin may prevent CHIKV infection because of its high binding affinity (−9.8 kcal/mol) for the CHIKV nsP3 protein. Lutein prevents HIV-1 reactivation by blocking clade B- and C-Tat-driven LTR transactivation [202–205] . By reducing the binding of the transcription factor Sp1, luteolin prevented the reactivation of the Epstein-Barr virus (EBV). discovered that luteolin interfered with viral RNA replication. In addition to these antiviral characteristics, luteolin or luteolin-rich fractions showed antiviral efectiveness against rotavirus, CHIKV, JEV, SARS-CoV, and other viruses [192, 206, 207]. The foundation of flavonol is 3-hydroxy-2-phenylchromen-4-one. Kaempferol and quercetin showed potential as antiviral agents. For instance, manager antiviral activity against the respiratory syncytial virus (RSV), HSV-1, and HSV-2 were improved by quercetin in dose-dependent ways in cell cultures [208, 209]. Flavans are defined by 2-phenyl-3,4-dihydro-2H-chromene. Flavan-3,4-diol, flavan-4, and flavan-3,3-ol. Epigallocatechin (EGC), catechin, epicatechin gallate, and epigallocatechin gallate (EGCG) are tea flavan-3-ols that have antiviral properties. Since Rawangkan et al. [210] showed that tea catechins, particularly EGCG, may bind to influenza virus haemagglutinin and prevent its adsorption to Madin-Darby canine kidney cells, the influenza virus has drawn the most attention as a potential target. Recent research demonstrates the antiviral potency of quercetin against several influenza virus strains. By interacting with influenza hemagglutinin, it prevents viral-cell fusion [193]. Quercetin inhibits transcription, protein synthesis, and viral endocytosis. Infected mice with airway cholinergic hyperresponsiveness and rhinovirus multiplication are both decreased by quercetin. HCMV is prevented by kaempferol. Coronavirus 3a channel blockers made of kaempferol derivatives with rhamnose residue are available. The influenza A virus is inhibited in vitro by kaempferol 3-O—L-rhamnopyranoside from Zanthoxylum piperitum. HIV-1 reverse transcriptase is inhibited by kaempferol and kaempferol-7-O-glucoside [106, 211–215]. The viral envelope may be damaged by EGCG, which prevents influenza virus and cellular membrane hemifusion [216–219]. Studies have shown that tea catechins are anti- HIV. Due to its ability to combat HIV-1 at all stages of its existence, EGCG is the most potent tea catechin. It inhibits gp120 binding by forming a direct bond with CD4 molecules. CD4's Trp69, Arg59, and Phe43 may interact with EGCG. Viral gp120 interacts with residues [192, 220]. By inhibiting the MEK/ERK1/2 and PI3-K/Akt signaling pathways, EGCG may shorten the EBV lytic cycle. Additionally, HCV cannot adhere to cells or replicate RNA when EGCG is present. The Zika virus (ZIKV) is inhibited by EGCG, according to a recent study: In vitro, the flavanone naringenin inhibits virus multiplication. The absence of anti-Sindbis virus activity in naringenin's glycoside form, naringin, demonstrates that rutinose restricts the antiviral action of this compound. Long-term treatment lowers HCV by 1.4 logs and may inhibit intracellular HCV particle assembly [221–224]. Treatment with naringenin after entrance reduced CHIKV in infected Vero cells. Bovine herpesvirus type 1 and New World

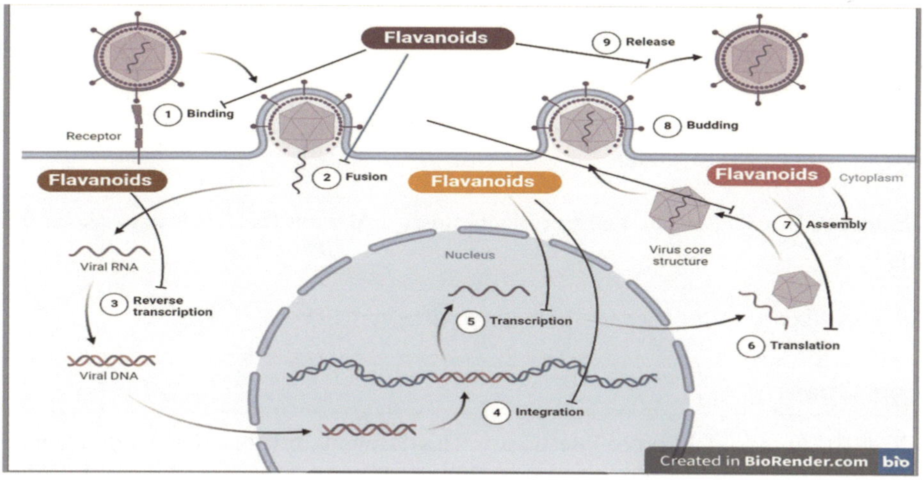

Figure 5.
Shows the protective effect of flavonoids against viral infection. Created with BioRender.com.

arenavirus Pichinde replication are decreased by the isoflavonoid genistein, which inhibits tyrosine kinase. Resting CD4 T cells and macrophages are protected against HIV infection by genistein. Avian leucosis virus subgroup J, HSV-1, and HSV-2 reproduction were all suppressed by genistein (**Figure 5**) [192, 225].

Author details

Asmaa Nabil-Adam[1*], Mohamed E. Elnosary[2], Mohamed L. Ashour[3,4], Nehad M. Abd El-Moneam[5] and Mohamed A. Shreadah[1]

1 Marine Biotechnology and Natural Products Lab (MBNP), National Institute of Oceanography and Fisheries (NIOF), Egypt

2 Faculty of Science, Botany and Microbiology Department, Al-Azhar University, Cairo, Egypt

3 Faculty of Pharmacy, Department of Pharmacognosy, Ain-Shams University, Cairo, Egypt

4 Department of Pharmaceutical Sciences, Pharmacy Program, Batterjee Medical College, Jeddah, Saudi Arabia

5 Faculty of Science, Department of Biochemistry, Alexandria University, Alexandria, Egypt

*Address all correspondence to: sama.biomarine@gmail.com

References

[1] Panche AN, Diwan AD, Chandra SR. Flavonoids: An overview. Journal of Nutritional Science. 2016;**5**:e47. DOI: 10.1017/jns.2016.41

[2] Santos EL, Maia BHLS, Teixeira APFD. Flavonoids: Classification, biosynthesis and chemical ecology. In: Flavonoids - from Biosynthesis to Human Health. London, UK, London, UK: IntechOpen; 2017. DOI: 10.5772/67861

[3] Mierziak J, Kostyn K, Kulma A. Flavonoids as important molecules of plant interactions with the environment. Molecules. 2014;**19**:16240-16265. DOI: 10.3390/molecules191016240 molecules ISSN 1420-3049. www.mdpi.com/journal/molecules

[4] Nijveldt RJ, Nood EV, van Hoorn DE, Boelens PG, van Norren K, van Leeuwen PA. Flavonoids: A review of probable mechanisms of action and potential applications. The American Journal of Clinical Nutrition. 2001; **74**(4):418-425. DOI: 10.1093/ajcn/74.4.418

[5] Alseekh S, de Souza LP, Benina M, Fernie AR. The style and substance of plant flavonoid decoration; towards defining both structure and function. Phytochemistry. 2020;**174**:112347

[6] Fujino N, Tenma N, Waki T, Ito K, Komatsuzaki Y, Sugiyama K, et al. Physical interactions among flavonoid enzymes in snapdragon and torenia reveal the diversity in the flavonoid metabolon organization of different plant species. The Plant Journal. 2018;**94**: 372-392

[7] Achnine L, Blancaflor EB, Rasmussen S, Dixon RA. Colocalization of L-phenylalanine ammonia-lyase and cinnamate 4-hydroxylase for metabolic channeling in phenylpropanoid biosynthesis. The Plant Cell. 2004; **16**(11):3098-3109. DOI: 10.1105/tpc.104.024406. Epub 2004 Oct 7

[8] Nakayama T, Takahashi S, Waki T. Formation of flavonoid metabolons: Functional significance of protein-protein interactions and impact on flavonoid Chemodiversity. Frontiers in Plant Science. 2019;**10**:821. DOI: 10.3389/fpls.2019.00821

[9] Ferreyra MLF, Rius SP, Casati P. Flavonoids: Biosynthesis, biological functions, and biotechnological applications. Frontiers in Plant Science. 2012;**3**:222. DOI: 10.3389/fpls.2012.00222

[10] Taboada C, Millan R, Miguez I. Evaluation of marine algae Undaria pinnatifida and Porphyra purpurea as a food supplement: Composition, nutritional value and effect of intake on intestinal, hepatic and renal enzyme activities in rats. Journal of the Science of Food and Agriculture. 2013;**93**(8):1863-1868. DOI: 10.1002/jsfa.5981. Epub 2013 Apr 16

[11] Kukil K, Lindberg P. Expression of phenylalanine ammonia lyases in Synechocystis sp. PCC 6803 and subsequent improvements of sustainable production of phenylpropanoids. Microbial Cell Factories. 2022;**21**(1):1-16. DOI: 10.1186/s12934-021-01735-8

[12] Rudrapal M, Khan J, Dukhyil AAB, Alarousy RMII, Attah EI, Sharma T, et al. Chalcone scaffolds, bioprecursors of flavonoids: Chemistry, bioactivities, and pharmacokinetics. Molecules. 2021; **26**(23):7177. DOI: 10.3390/molecules26237177

[13] Khare S, Dewangan R, Kumar A. Structure-Activity Relationship of

Flavonoids. In: Chaurasia PK, Bharati SL, editors. The Chemistry Inside Spices and Herbs: Research and Development. Bentham Science Publisher; 2022. pp. 235-256. DOI: 10.2174/9789815039566122010011.

[14] Dixon RA, Pasinetti GM. Flavonoids and isoflavonoids: From plant biology to agriculture and neuroscience. Plant Physiology. 2010;**154**(2):453-457. DOI: 10.1104/pp.110.161430

[15] Brglez Mojzer E, Knez Hrnčič M, Škerget M, Knez Ž, Bren U. Polyphenols: Extraction methods, Antioxidative action, bioavailability and Anticarcinogenic effects. Molecules. 2016;**21**(7):901. DOI: 10.3390/molecules21070901

[16] Shrestha S, Smid S, Zhang W. Phlorotannins: A review on biosynthesis, chemistry and bioactivity. Food Bioscience. 2020;**27**(14): R713-R715. DOI: 10.1016/j.fbio.2020.100832

[17] Tziveleka LA, Tammam MA, Tzakou O, Roussis V, Ioannou E. Metabolites with antioxidant activity from marine macroalgae. Antioxidants (Basel). 2021;**10**(9):1431. DOI: 10.3390/antiox10091431

[18] Spiegel M, Sroka Z, Andruniów T. Flavones' and Flavonols' antiradical structure–activity relationship—A quantum chemical study. Antioxidants. 2020;**9**:461. DOI: 10.3390/antiox9060461

[19] Ullah A, Munir S, Badshah SL, Khan N, Ghani L, Poulson BG, et al. Important flavonoids and their role as a therapeutic agent. Molecules. 2020; **25**(22):5243. DOI: 10.3390/molecules25225243

[20] Khoo HE, Azlan A, Tang ST, Lim SM. Anthocyanidins and anthocyanins: Colored pigments as food, pharmaceutical ingredients, and the potential health benefits. Food & Nutrition Research. 2017;**61**(1):1361779. DOI: 10.1080/16546628.2017.1361779

[21] Salehi B, Quispe C, Chamkhi I, El Omari N, Balahbib A, Sharifi-Rad J, et al. Pharmacological properties of Chalcones: A review of preclinical including molecular mechanisms and clinical evidence. Frontiers in Pharmacology. 2021;**11**:592654. DOI: 10.3389/fphar.2020.592654

[22] Wang L, Yang B, Du X, Yi C. Optimisation of supercritical fluid extraction of flavonoids from Pueraria lobata. Food Chemistry. 2008;**108**(2): 737-741. DOI: 10.1016/j.foodchem.2007.11.031. Epub 2007 Nov 22

[23] Jiang N, Doseff AI, Grotewold E. Flavones: From biosynthesis to health benefits. Plants. 2016;**5**:27. DOI: 10.3390/plants5020027

[24] Reiter E, Gerster P, Jungbauer A. Red clover and soy isoflavones-an in vitro safety assessment. Gynecological Endocrinology: the Official Journal of the International Society of Gynecological Endocrinology. 2011;**27**:1037-1042. DOI: 10.3109/09513590.2011.588743

[25] Makhaik MS, Shakya AK, Kale R. Dietary phytochemicals: As a natural source of antioxidants. In: Antioxidants - Benefits, Sources, Mechanisms of Action. London, UK, London, UK: IntechOpen; 2021. DOI: 10.5772/intechopen.99159

[26] Khalid M, Saeed-ur-R QM, Huang D-f. Role of flavonoids in plant interactions with the environment and against human pathogens — A review. Journal of Integrative Agriculture. 2019; **18**:211-230. DOI: 10.1016/S2095-3119(19)62555-4

[27] Samanta A, Das G, Das S. Roles of flavonoids in plants. International Journal of Pharmaceutical Science and Technology. 2011;**6**:12-35

[28] Imamovic B, Trebse P, Omeragic E, Becic E, Pecet A, Dedic M. Stability and removal of benzophenone-type UV filters from water matrices by advanced oxidation processes. Molecules. 2022;**27**: 1874. DOI: 10.3390/molecules27061874

[29] Laoué J, Fernandez C, Ormeño E. Plant flavonoids in Mediterranean species: A focus on Flavonols as protective metabolites under climate stress. Plants. 2022;**11**:172. DOI: 10.3390/plants11020172

[30] Amawi H, Ashby CR, Tiwari AK. Cancer chemoprevention through dietary flavonoids: what's limiting? Chinese Journal of Cancer. 2017;**36**:50. DOI: 10.1186/s40880-017-0217-4

[31] Kozłowska A, Szostak-Węgierek D. Flavonoids – Food sources, health benefits, and mechanisms involved. In: Mérillon JM, Ramawat K, editors. Bioactive Molecules in Food. Reference Series in Phytochemistry. Cham: Springer; 2018. DOI: 10.1007/978-3-319-54528-8_54-1

[32] Rodriguez MC, Caleja C, Nuñez-Estevez B, Pereira E, Fraga-Corral M, Reis FS, et al. Flavonoids: A Group of Potential Food Additives with beneficial health effects. In: Lage MÁÁP, Otero P, editors. Natural Food Additives [Working Title]. London, UK, London, UK: IntechOpen; 2021. DOI: 10.5772/intechopen.101466

[33] Alghazeer R, Howell N, El-Naili M, Awayn N. Anticancer and antioxidant activities of some algae from Western Libyan coast. Natural Science. 2018;**10**:232-246. DOI: 10.4236/ns.2018.107025

[34] Bonfante P. Algae and fungi move from the past to the future. eLife. 2019;**8**: e49448. DOI: 10.7554/eLife.49448

[35] Nascimento LB, Tattini M. Beyond Photoprotection: The multifarious roles of flavonoids in plant Terrestrialization. International Journal of Molecular Sciences. 2022;**23**(9):5284. DOI: 10.3390/ijms23095284

[36] Goiris K, Muylaert K, Voorspoels S, Noten B, De Paepe D, Baart G, et al. Detection of flavonoids in microalgae from different evolutionary lineages. Journal of Phycology. 2014;**50**:483-492. DOI: 10.1111/jpy.12180

[37] Bag S, Mondal A, Mondal SK MA, Banik A. Flavonoid mediated selective cross-talk between plants and beneficial soil microbiome. Phytochemistry Reviews. 2022;1-22. DOI: 10.1007/s11101-022-09806-3

[38] Bais H, Walker T, Kennan A, Stermitz F, Vivanco J. Structure-dependent Phytotoxicity of Catechins and other flavonoids: Flavonoid conversions by cell-free protein extracts of Centaurea maculosa (spotted knapweed) roots. Journal of Agricultural and Food Chemistry. 2003;**51**:897-901. DOI: 10.1021/jf020978a

[39] Cheng F, Zi C. Research Progress on the use of plant allelopathy in agriculture and the physiological and ecological mechanisms of allelopathy. Frontiers in Plant Science. 2015;**6**:1020. DOI: 10.3389/fpls.2015.01020

[40] Drew MC, He CJ, Morgan PW. Programmed cell death and aerenchyma formation in roots. Trends in Plant Science. 2000;**5**(3):123-127. DOI: 10.1016/s1360-1385(00)01570-3

[41] Kim TY, Leem E, Lee JM, Kim SR. Control of reactive oxygen species for

the prevention of Parkinson's disease: The possible application of flavonoids. Antioxidants (Basel). 2020;**9**(7):583. DOI: 10.3390/antiox9070583

[42] Kumar S, Pandey AK. Chemistry and biological activities of flavonoids: An overview. Scientific World Journal. 2013;**162750**. DOI: 10.1155/2013/162750

[43] Cassedy A, Parle-McDermott A, O'Kennedy R. Virus detection: A review of the current and emerging molecular and immunological methods. Frontiers in Molecular Biosciences. 2021;**8**:637559. DOI: 10.3389/fmolb.2021.637559

[44] Koirala N, Thuan N, Ghimire G, Thang D, Sohng J-K. Methylation of flavonoids: Chemical structures, bioactivities, progress and perspectives for biotechnological production. Enzyme and Microbial Technology. 2016;**86**:103-116. DOI: 10.1016/j.enzmictec.2016.02.003

[45] Förster C, Handrick V, Ding Y, Nakamura Y, Paetz C, Schneider B, et al. Biosynthesis and antifungal activity of fungus-induced *O*-methylated flavonoids in maize. Plant Physiology. 2022;**188**(1):167-190. DOI: 10.1093/plphys/kiab496

[46] Rupasinghe H, Vasantha P. Special issue "flavonoids and their disease prevention and treatment potential": Recent advances and future perspectives. Molecules. 2020;**25**:4746. DOI: 10.3390/molecules25204746

[47] Mondal S, Rahaman ST. Flavonoids: A vital resource in healthcare and medicine. Pharmacy & Pharmacology International Journal. 2020;**8**(2):91-104. DOI: 10.15406/ppij.2020.08.00285

[48] Stan D, Enciu AM, Mateescu AL, Ion AC, Brezeanu AC, Stan D, et al. Natural compounds with antimicrobial and antiviral effect and Nanocarriers used for their transportation. Frontiers in Pharmacology. 2021;**12**:723233. DOI: 10.3389/fphar.2021.723233

[49] Ninfali P, Antonelli A, Magnani M, Scarpa E. Antiviral properties of flavonoids and delivery strategies. Nutrients. 2020;**12**:2534. DOI: 10.3390/nu12092534

[50] Wang L, Pan X, Jiang L, Chu Y, Gao S, Jiang X, et al. The biological activity mechanism of Chlorogenic acid and its applications in food industry: A review. Frontiers in Nutrition. 2022;**9**: 943911. DOI: 10.3389/fnut.2022.943911

[51] Conteh E, Okereke M, Turay FU, Bah AS, Muhsinah A. The need for a functional pharmaceutical industry in Sierra Leone: Lessons from the COVID-19 pandemic. Journal of Pharmacy Policy and Practice. 2022;**15**:46. DOI: 10.1186/s40545-022-00444-w

[52] Ramešová Š, Sokolová R, Degano I, Bulíčková J, Žabka J, Gál M. On the stability of the bioactive flavonoids quercetin and luteolin under oxygen-free conditions. Analytical and Bioanalytical Chemistry. 2011;**402**:975-982. DOI: 10.1007/s00216-011-5504-3

[53] Veiko AG, Lapshina EA, Zavodnik IB. Comparative analysis of molecular properties and reactions with oxidants for quercetin, catechin, and naringenin. Molecular and Cellular Biochemistry. 2021;**476**(12):4287-4299. DOI: 10.1007/s11010-021-04243-w

[54] Xie L, Deng Z, Zhang J, Dong H, Wang W, Xing B, et al. Comparison of flavonoid O-glycoside, C-glycoside and their Aglycones on antioxidant capacity and metabolism during In vitro digestion and In vivo. Food. 2022;**11**(6):882. DOI: 10.3390/foods11060882

[55] Conesa C, Yanez J, Vicente V, Alcaraz M, Benavente-García O, Castillo J, et al. Effects of several polyhydroxylated flavonoids on the growth of B16F10 melanoma and Melan-a melanocyte cell lines: Influence of the sequential oxidation state of the flavonoid skeleton. Melanoma Research. 2003;**13**:3-9. DOI: 10.1097/01.cmr.0000043160.28051.64

[56] Kopustinskiene DM, Jakstas V, Savickas A, Bernatoniene J. Flavonoids as anticancer agents. Nutrients. 2020;**12**(2):457. DOI: 10.3390/nu12020457

[57] Naksuriya O, Daowtak K, Tima S, Okonogi S, Mueller M, Toegel S, et al. Hydrolyzed flavonoids from *Cyrtosperma johnstonii* with superior antioxidant, Antiproliferative, and anti-inflammatory potential for cancer prevention. Molecules. 2022;**27**(10):3226. DOI: 10.3390/molecules27103226

[58] Katalinić M, Rusak G, Barović J, Sinko G, Jelic D, Kovarik Z. Structural aspects of flavonoids as inhibitors of human butyrylcholinesterase. European Journal of Medicinal Chemistry. 2009; **45**:186-192. DOI: 10.1016/j.ejmech.2009.09.041

[59] Dabravolski SA, Sukhorukov VN, Kalmykov VA, Grechko AV, Shakhpazyan NK, Orekhov AN. The role of KLF2 in the regulation of atherosclerosis development and potential use of KLF2-targeted therapy. Biomedicine. 2022;**10**(2):254. DOI: 10.3390/biomedicines10020254

[60] Martínez L, Pons Vila Z, Margalef M, Arola-Arnal A, Muguerza B. Regulation of vascular endothelial genes by dietary flavonoids: Structure-expression relationship studies and the role of the transcription factor KLF-2. The Journal of Nutritional Biochemistry. 2015;**26**:277-284. DOI: 10.1016/j.jnutbio.2014.11.003

[61] Caporali S, De Stefano A, Calabrese C, Giovannelli A, Pieri M, Savini I, et al. Anti-inflammatory and active biological properties of the plant-derived bioactive compounds Luteolin and Luteolin 7-glucoside. Nutrients. 2022;**14**(6):1155. DOI: 10.3390/nu14061155

[62] Harmon A, Harp J. Differential effects of flavonoids on 3T3-L1 adipogenesis and lipolysis. American Journal of Physiology-Cell Physiology. 2001;**280**:C807-C813. DOI: 10.1152/ajpcell.2001.280.4.C807

[63] Dinda B, Dinda M, Roy A, Dinda S. Dietary plant flavonoids in prevention of obesity and diabetes. Advances in Protein Chemistry and Structural Biology. 2020;**120**:159-235. DOI: 10.1016/bs.apcsb.2019.08.006

[64] Al-Khayri JM, Sahana GR, Nagella P, Joseph BV, Alessa FM, Al-Mssallem MQ. Flavonoids as potential anti-Inflflammatory molecules: A review. Molecules. 2022;**27**:2901. DOI: 10.3390/molecules27092901

[65] Górniak I, Bartoszewski R, Króliczewski J. Comprehensive review of antimicrobial activities of plant flavonoids. Phytochemistry Reviews. 2019;**18**:241-272. DOI: 10.1007/s11101-018-9591-z

[66] Rolt A, Cox LS. Structural basis of the anti-ageing effects of polyphenolics: Mitigation of oxidative stress. BMC Chemistry. 2020;**14**:50. DOI: 10.1186/s13065-020-00696-0

[67] Pannala A, Chan T, O'Brien P, Rice-Evans C. Flavonoid B-ring chemistry and antioxidant activity: Fast reaction kinetics. Biochemical and Biophysical

Research Communications. 2001;**282**: 1161-1168. DOI: 10.1006/bbrc.2001.4705

[68] Musialik M, Kuzmicz R, Pawłowski T, Litwinienko G. Acidity of hydroxyl groups: An overlooked influence on antiradical properties of flavonoids. The Journal of Organic Chemistry. 2009;**74**:2699-2709. DOI: 10.1021/jo802716v

[69] Al-Mamary MA, Moussa Z. Antioxidant activity: The presence and impact of hydroxyl groups in small molecules of natural and synthetic origin. In: Antioxidants - Benefits, Sources, Mechanisms of Action. London, UK, London, UK: IntechOpen; 2021. DOI: 10.5772/intechopen.95616

[70] Platzer M, Kiese S, Tybussek T, Herfellner T, Schneider F, Schweiggert-Weisz U, et al. Radical scavenging mechanisms of phenolic compounds: A quantitative structure-property relationship (QSPR) study. Frontiers in Nutrition. 2022;**9**:882458. DOI: 10.3389/fnut.2022.882458

[71] Sánchez-Marzo N, Pérez-Sánchez A, Ruiz-Torres V, Martínez-Tébar A, Castillo J, Herranz Lopez M, et al. Antioxidant and Photoprotective activity of Apigenin and its potassium salt derivative in human keratinocytes and absorption in Caco-2 cell monolayers. International Journal of Molecular Sciences. 2019;**20**:2148. DOI: 10.3390/ijms20092148

[72] Barreca D, Bisignano C, Ginestra G, Bisignano G, Bellocco E, Leuzzi U, et al. Polymethoxylated, C- and O-glycosyl flavonoids in tangelo (*C. reticulata* × *C. paradisi*) juice and their influence on antioxidant properties. Food Chemistry. 2013;**141**(2):1481-1488. DOI: 10.1016/j.foodchem.2013.03.095

[73] Ekalu A, Habila JD. Flavonoids: Isolation, characterization, and health benefits. Beni-Suef University Journal of Basic and Applied Sciences. 2020;**9**:45. DOI: 10.1186/s43088-020-00065-9

[74] Xiao J, Chen T, Cao H, Chen L, Yang F. Molecular property-affinity relationship of flavanoids and flavonoids for HSA in vitro. Molecular Nutrition & Food Research. 2011;**55**:310-317. DOI: 10.1002/mnfr.201000208

[75] Lewandowski W, Lewandowska H, Golonko A, Świderski G, Świsłocka R, Kalinowska M. Correlations between molecular structure and biological activity in "logical series" of dietary chromone derivatives. PLoS One. 2020;**15**(8):e0229477. DOI: 10.1371/journal.pone.0229477

[76] Sarian MN, Ahmed QU, Mat So'ad SZ, Alhassan AM, Murugesu S, Perumal V, et al. Antioxidant and antidiabetic effects of flavonoids: A structure-activity relationship based study. BioMed Research International. 2017;**2017**(2017):8386065. DOI: 10.1155/2017/8386065. Epub 2017 Nov 28

[77] Li C, Dai T, Chen J, Chen M, Liang R, Liu C, et al. Modification of flavonoids: Methods and influences on biological activities. Critical Reviews in Food Science and Nutrition. 2022;1-22. DOI: 10.1080/10408398.2022.2083572

[78] Kaurinovic B, Vastag D. Flavonoids and phenolic acids as potential natural antioxidants. In: Antioxidants. London, UK, London, UK: IntechOpen; 2019. DOI: 10.5772/intechopen.83731

[79] Xiao J, Capanoglu E, Jassbi AR, Miron A. Advance on the flavonoid *C*-glycosides and health benefits. Critical Reviews in Food Science and Nutrition. 2016;**56**(sup. 1):S29-S45. DOI: 10.1080/10408398.2015.1067595

[80] Manach C, Donovan J. Pharmacokinetics and metabolism of dietary flavonoids in humans. Free Radical Research. 2004;**38**:771-785

[81] El-Kattan A, Varma M. Oral absorption, intestinal metabolism and human Oral bioavailability. In: Topics on Drug Metabolism. London, UK, London, UK: IntechOpen; 2012. DOI: 10.5772/31087

[82] Cassidy A, Minihane A-M. The role of metabolism (and the microbiome) in defining the clinical efficacy of dietary flavonoids. The American Journal of Clinical Nutrition. 2017;**105**(1):10-22. DOI: 10.3945/ajcn.116.136051

[83] Murota K, Nakamura Y, Uehara M. Flavonoid metabolism: The interaction of metabolites and gut microbiota. Bioscience, Biotechnology, and Biochemistry. 2018;**82**:4, 600-610. DOI: 10.1080/09168451.2018.1444467

[84] Hussain MB, Hassan S, Waheed M, Javed A, Farooq MA, Tahir A. Bioavailability and metabolic pathway of phenolic compounds. In: Soto-Hernández M, García-Mateos R, Palma-Tenango M, editors. Plant Physiological Aspects of Phenolic Compounds. London, UK, London, UK: IntechOpen; 2019. DOI: 10.5772/intechopen.84745

[85] Chen X, Yin O, Zuo Z, Chow M. Pharmacokinetics and modeling of quercetin and metabolites. Pharmaceutical Research. 2005;**22**: 892-901. DOI: 10.1007/s11095-005-4584-1

[86] Tiho T, Yao J, Brou Y, Adima A. Drying temperature effect on Total phenols and flavonoids content, and antioxidant activity of Borassus aethiopum Mart ripe fruits' Pulp. Journal of Food Research. 2017;**6**:50. DOI: 10.5539/jfr.v6n2p50

[87] Zeng Q, Xu Z, Dai M, Cao X, Xiong X, He S, et al. Effects of simulated digestion on the phenolic composition and antioxidant activity of different cultivars of lychee pericarp. BMC Chemistry. 2019;**13**:27. DOI: 10.1186/s13065-019-0544-4

[88] Dias MC, Pinto DCGA, Silva AMS. Plant flavonoids: Chemical characteristics and biological activity. Molecules. 2021;**26**(17):5377. DOI: 10.3390/molecules26175377

[89] Sajid M, Channakesavula CN, Stone SR, Kaur P. Synthetic biology towards improved flavonoid pharmacokinetics. Biomolecules. 2021; **11**(5):754. DOI: 10.3390/biom11050754

[90] Viskupicova J, Ondrejovič M, Sturdik E. Bioavailability and metabolism of flavonoids. Journal of Food and Nutrition Research. 2008;**47**: 151-162

[91] Shreadah MA, Abdel Monein NM, El-Assar SA, Nabil-Adam A. Phytochemical and pharmacological screening of Sargassium vulgare from Suez Canal, Egypt. Food Science and Biotechnology. 2018a;**27**(4):963-979

[92] Shreadah MA, Abou Ella HM, Abdel Monein NM, Yakout GA. Isolation, phylogenetic analysis of the microbial community associated with the Red Sea sponge Ircinia Echinata and biological evaluation of their secondary metabolites. Biomedical Journal of Scientific & Technical Research, Biomedical Research Network+, LLC. 2018b;**12**(2):9064-9082

[93] Shreadah MA, Abdel Monein NM, Abou Ella HM. Bacteria from marine sponges: A source of biologically active compounds. Biomedical Journal of Scientific & Technical Research. 2018c; **10**(5):1-20

[94] Lopes AFC. Mitochondrial metabolism and DNA methylation: A review of the interaction between two genomes. Clinical Epigenetics. 2020;**12**: 182. DOI: 10.1186/s13148-020-00976-5

[95] Zou X, Liu Y-C, Hsu N-S, Huang C-J, Lyu S-Y, Chan H, et al. Structure and mechanism of a nonhaem-iron SAM-dependent C-methyltransferase and its engineering to a hydratase and an O-methyltransferase. Acta Crystallographica. Section D, Biological Crystallography. 2014;**70**:1549-1560. DOI: 10.1107/S1399004714005239

[96] Lee S, Kang J, Kim J. Structural and biochemical characterization of Rv0187, an O-methyltransferase from *mycobacterium tuberculosis*. Scientific Reports. 2019;**9**:8059. DOI: 10.1038/s41598-019-44592-7

[97] Yang B, Liu H, Jiali Y, Gupta V, Jiang Y. New insights on bioactivities and biosynthesis of flavonoid glycosides. Trends in Food Science & Technology. 2018;**141**(2):1481-1488. DOI: 10.1016/j.tifs.2018.07.006

[98] Slámová K, Kapešová J, Valentová K. "sweet flavonoids": Glycosidase-catalyzed modifications. International Journal of Molecular Sciences. 2018; **19**(7):2126. DOI: 10.3390/ijms19072126

[99] Stromsnes K, Lagzdina R, Olaso-Gonzalez G, Gimeno-Mallench L, Gambini J. Pharmacological properties of polyphenols: Bioavailability, mechanisms of action, and biological effects in In vitro studies, animal models, and humans. Biomedicine. 2021;**9**(8): 1074. DOI: 10.3390/biomedicines9081074

[100] Fabbrini M, D'Amico F, Barone M, Conti G, Mengoli M, Brigidi P, et al. Polyphenol and tannin nutraceuticals and their metabolites: How the human gut microbiota influences their properties. Biomolecules. 2022;**12**(7):875. DOI: 10.3390/biom12070875

[101] Nabil-Adam A, Shreadah MA, Abdel Monein NM, El-Assar SA. Pesudomance sp. bacteria associated with marine sponge as a promising and sustainable source of bioactive molecules. Current Pharmaceutical Biotechnology. 2019;**20**(11):964-984. DOI: 10.2174/1389201020666190619092502

[102] Nabil-Adam A, Shreadah MA, Abd El-Moneam NM, El-Assar SA. Various In vitro bioactivities of secondary metabolites isolated from the sponge Hyrtios aff. Erectus from the Red Sea coast of Egypt. Turkish Journal of Pharmaceutical Sciences. 2020a;**17**(2):127-135. DOI: 10.4274/tjps.galenos.2018.72677

[103] Nabil-Adam A, Shreadah MA, Abd El-Moneam NM, El-Assar SA. Marine algae of the genus Gracilaria as multi products source for different biotechnological and medical applications. Recent Patents on Biotechnology. 2020b;**14**(3):203-228. DOI: 10.2174/1872208314666200121144816

[104] Nabil-Adam A, Shreadah MA. Red algae natural products for prevention of lipopolysaccharides (LPS)- induced liver and kidney inflammation and injuries. Bioscience Reports. 2021a;**41**(1): BSR20202022. DOI: 10.1042/BSR20202022

[105] Nabil-Adam A, Shreadah MA. Ameliorative role of Ulva extract against heavy metal mixture—Induced cardiovascular through oxidative/antioxidant pathways and inflammatory biomarkers. Environmental Science and Pollution Research International. 2021b; **28**(21):27006-27024. DOI: 10.1007/s11356-020-11994-4. Epub 2021 Jan 26

[106] Nabil-Adam A, Shreadah MA. Anti-inflammatory, antioxidant, lung and liver protective activity of Galaxaura oblongata as antagonistic efficacy against LPS using hematological parameters and immunohistochemistry as biomarkers. Cardiovascular & Hematological Agents in Medicinal Chemistry. 2022;**20**(2): 148-165. DOI: 10.2174/1871525719666210112154800. Epub ahead of print

[107] Ciumărnean L, Milaciu MV, Runcan O, Vesa ȘC, Răchișan AL, Negrean V, et al. The effects of flavonoids in cardiovascular diseases. Molecules. 2020;**25**(18):4320. DOI: 10.3390/molecules25184320

[108] Grijalva-Guiza RE, Jiménez-Garduño AM, Hernández LR. Potential benefits of flavonoids on the progression of atherosclerosis by their effect on vascular smooth muscle excitability. Molecules. 2021;**26**(12):3557. DOI: 10.3390/molecules26123557

[109] Mohd Zaid NA, Sekar M, Bonam SR, Gan SH, Lum PT, Begum MY, et al. Promising natural products in new drug design, development, and therapy for skin disorders: An overview of scientific evidence and understanding their mechanism of action. Drug Design, Development and Therapy. 2022;**16**: 23-66. DOI: 10.2147/DDDT.S326332

[110] Batiha GE, Beshbishy AM, Ikram M, Mulla ZS, El-Hack MEA, Taha AE, et al. The pharmacological activity, biochemical properties, and pharmacokinetics of the major natural polyphenolic flavonoid: Quercetin. Food. 2020;**9**(3):374. DOI: 10.3390/foods9030374

[111] El-Darier SM, Rashed SA, Fayez A, Hassanein SS, Sharaby MR, Tawfik NM, et al. Medicinal plant-derived compounds as potential phytotherapy for COVID-19: Future perspectives. Journal of Pharmacognosy and Phytotherapy. 2021;**13**(3):68-81

[112] Mendes V, Vilaça R, Freitas V, Ferreira P, Mateus N, Costa V. Effect of Myricetin, Pyrogallol, and Phloroglucinol on yeast resistance to oxidative stress. Oxidative Medicine and Cellular Longevity. 2015;**2015**:1-10. DOI: 10.1155/2015/782504

[113] Gil JV, Esteban-Muñoz A, Fernández-Espinar MT. Changes in the polyphenolic profile and antioxidant activity of wheat bread after incorporating quinoa flour. Antioxidants. 2022;**11**:33. DOI: 10.3390/antiox11010033

[114] Imran M, Salehi B, Sharifi-Rad J, Aslam Gondal T, Saeed F, Imran A, et al. Kaempferol: A key emphasis to its anticancer potential. Molecules. 2019; **24**(12):2277. DOI: 10.3390/molecules24122277

[115] Sharifi-Rad J, Cruz-Martins N, López-Jornet P, Lopez EP, Harun N, Yeskaliyeva B, et al. Natural Coumarins: Exploring the pharmacological complexity and underlying molecular mechanisms. Oxidative Medicine and Cellular Longevity. 2021;**2021**:6492346. DOI: 10.1155/2021/6492346

[116] Kiokias S, Proestos C, Oreopoulou V. Phenolic acids of plant origin-a review on their antioxidant activity In vitro (O/W emulsion systems) along with their in vivo health biochemical properties. Food. 2020;**9**(4):534. DOI: 10.3390/foods9040534

[117] Sen A, Atmaca P, Terzioglu G, Arslan S. Anticarcinogenic effect and carcinogenic potential of the dietary phenolic acid: o-coumaric acid. Natural

Product Communications. 2013;**8**(9): 1269-1274

[118] Sova M, Saso L. Natural sources, pharmacokinetics, biological activities and health benefits of Hydroxycinnamic acids and their metabolites. Nutrients. 2020;**12**(8):2190. DOI: 10.3390/nu12082190

[119] Taïlé J, Bringart M, Planesse C, Patché J, Rondeau P, Veeren B, et al. Antioxidant polyphenols of *Antirhea borbonica* medicinal plant and Caffeic acid reduce cerebrovascular, inflammatory and metabolic disorders aggravated by high-fat diet-induced obesity in a mouse model of stroke. Antioxidants (Basel). 2022;**11**(5):858. DOI: 10.3390/antiox11050858

[120] Xu T, Song Q, Zhou L, Yang W, Wu X, Qian Q, et al. Ferulic acid alleviates lipotoxicity-induced hepatocellular death through the SIRT1-regulated autophagy pathway and independently of AMPK and Akt in AML-12 hepatocytes. Nutrition & Metabolism (London). 2021;**18**(1):13. DOI: 10.1186/s12986-021-00540-9

[121] Pizzino G, Irrera N, Cucinotta M, Pallio G, Mannino F, Arcoraci V, et al. Oxidative stress: Harms and benefits for human health. Oxidative Medicine and Cellular Longevity. 2017;**2017**:8416763. DOI: 10.1155/2017/8416763. Epub 2017 Jul 27

[122] Aboul-Ela HM, Shreadah MA, Abdel-Monem NM, Yakout GA, Van Soest RWM. Isolation, cytotoxic activity and phylogenetic analysis of bacillus sp. bacteria associated with the red sea sponge Amphimedonochracea. Advances in Bioscience and Biotechnology. 2012;**3**(7):815-823

[123] Abdel-Monem N, Abdel-Azeem AM, El Ashry ESH, Ghareeb DA, Adam AN. Assessment of secondary metabolites from marine-derived fungi as antioxidant. Journal of Medicinal Chemistry. 2013;**3**(3):60-73

[124] Abd El Moneam NM, Shreadah MA, Assar SA, Nabil-Adam A. Protective role of antioxidants capacity of Hyrtios aff. Erectus sponge extract against mixture of persistent organic pollutants (POPs)-induced hepatic toxicity in mice liver: Biomarkers and ultrastructural study. Environmental Science and Pollution Research. 2017;**24**(27):22061-22072

[125] Abd El Moneam NM, Yacout GA, Aboul-Ela HM, Shreadah MA. Hepatoprotective activity of chitosan Nanocarriers loaded with the ethyl acetate extract of Astenotrophomonas sp. bacteria associated with the Red Sea sponge Amphimedon Ochracea In CCl4 induced hepatotoxicity in rats. Advances in Bioscience and Biotechnology (ABB). 2017b;**8**(1):27-50

[126] Shreadah MA, Abdel Monein NM, El-Assar SA, Nabil-Adam A. The ameliorative role of a marine sponge extract against mixture of persistent organic pollutants induced changes in hematological parameters in mice. Expert Opinion on Environment Biology. 2017;**6**:2. DOI: 10.4172/2325-9655.1000143

[127] Shreadah MA, Abou Ella M, Abdel Monein NM, Yakout GA. Sponge associated bacteria- isolation, phylogentic Analysiand Biothnological potential. Biomedical Journal of Scientific Research, Biomedical Research Network+, LLC. 2019;**15**(2):11269-11285. DOI: 10.26717/BJSTR. 2019.15.002692

[128] Shreadah MA, Abd El-Moneam NM, El-Assar SA, Nabil-Adam A. Metabolomics and pharmacological screening of aspergillus versicolor isolated from *Hyrtios erectus* Red Sea

sponge; Egypt. Current Bioactive Compounds. 2020;**16**(1083): 1573407215666191111122711

[129] Tan BL, Norhaizan ME, Liew WP, Sulaiman RH. Antioxidant and oxidative stress: A mutual interplay in age-related diseases. Frontiers in Pharmacology. 2018;**9**:1162. DOI: 10.3389/fphar.2018.01162

[130] Meng X, Munishkina LA, Fink AL, Uversky VN. Effects of various flavonoids on the α-Synuclein fibrillation process. Parkinsons Disease. 2010;**2010**: 650794. DOI: 10.4061/2010/650794

[131] Lalhminghlui K, Jagetia GC. Evaluation of the free-radical scavenging and antioxidant activities of Chilauni, *Schima wallichii* Korth *in vitro*. Future Science OA. 2018;**4**(2):FSO272. DOI: 10.4155/fsoa-2017-0086

[132] Kurek-Górecka A, Rzepecka-Stojko A, Górecki M, Stojko J, Sosada M, Swierczek-Zieba G. Structure and antioxidant activity of polyphenols derived from propolis. Molecules. 2013; **19**(1):78-101. DOI: 10.3390/molecules19010078

[133] Chobot V, Hadacek F, Bachmann G, Weckwerth W, Kubicova L. Pro- and antioxidant activity of three selected Flavan type flavonoids: Catechin, Eriodictyol and Taxifolin. International Journal of Molecular Sciences. 2016;**17**: 1986. DOI: 10.3390/ijms17121986

[134] Cherrak SA, Mokhtari-Soulimane N, Berroukeche F, Bensenane B, Cherbonnel A, Merzouk H, et al. In vitro antioxidant versus metal Ion chelating properties of flavonoids: A structure-activity investigation. PLoS One. 2016; **11**(10):e0165575. DOI: 10.1371/journal.pone.0165575

[135] Galicia-Garcia U, Benito-Vicente A, Jebari S, Larrea-Sebal A, Siddiqi H, Uribe KB, et al. Pathophysiology of type 2 diabetes mellitus. International Journal of Molecular Sciences. 2020;**21**(17):6275. DOI: 10.3390/ijms21176275

[136] Lamport D. The effects of flavonoid and other polyphenol consumption on cognitive function: A systematic review of human experimental and epidemiological studies. Nutrtion and Aging. 2012;**1**:5-25. DOI: 10.3233/NUA-2012-0002

[137] Hole KL, Williams RJ. Flavonoids as an intervention for Alzheimer's disease: Progress and hurdles towards defining a mechanism of action. Brain Plasticity. 2021;**6**(2):167-192. DOI: 10.3233/BPL-200098

[138] Ayaz M, Sadiq A, Junaid M, Ullah F, Ovais M, Ullah I, et al. Flavonoids as prospective Neuroprotectants and their therapeutic propensity in aging associated neurological disorders. Frontiers in Aging Neuroscience. 2019; **11**:155. DOI: 10.3389/fnagi.2019.00155

[139] Onozuka H, Nakajima A, Matsuzaki K, Shin RW, Ogino K, Saigusa D, et al. Nobiletin, a citrus flavonoid, improves memory impairment and Abeta pathology in a transgenic mouse model of Alzheimer's disease. The Journal of Pharmacology and Experimental Therapeutics. 2008; **326**(3):739-744. DOI: 10.1124/jpet.108.140293. Epub 2008 Jun 10

[140] Uddin MS, Kabir MT, Niaz K, Jeandet P, Clément C, Mathew B, et al. Molecular insight into the therapeutic promise of flavonoids against Alzheimer's disease. Molecules. 2020; **25**(6):1267. DOI: 10.3390/molecules25061267

[141] Wang Y-J, Thomas P, Zhong J-H, Bi F-F, Kosaraju S, Pollard A, et al. Consumption of grape seed extract

prevents amyloid-β deposition and attenuates inflammation in brain of an Alzheimer's disease mouse. Neurotoxicity Research. 2009;**15**:3-14. DOI: 10.1007/s12640-009-9000-x

[142] Frandsen J, Narayanasamy P. Flavonoid enhances the glyoxalase pathway in cerebellar neurons to retain cellular functions. Scientific Reports. 2017;7:5126. DOI: 10.1038/s41598-017-05287-z

[143] Fernandes L, Cardim-Pires TR, Foguel D, Palhano FL. Green tea polyphenol epigallocatechin-Gallate in amyloid aggregation and neurodegenerative diseases. Frontiers in Neuroscience. 2021;**15**:718188. DOI: 10.3389/fnins.2021.718188

[144] Ehrnhoefer D, Bieschke J, Boeddrich A, Herbst M, Masino L, Lurz R, et al. EGCG redirects amyloidogenic polypeptides into unstructured, off-pathway oligomers. Nature Structural & Molecular Biology. 2008;**15**:558-566. DOI: 10.1038/nsmb.1437

[145] Pluta R, Januszewski S, Czuczwar SJ. Myricetin as a promising molecule for the treatment of post-ischemic brain neurodegeneration. Nutrients. 2021;**13**(2):342. DOI: 10.3390/nu13020342

[146] Chandrasekara A, Shahidi F. Herbal beverages: Bioactive compounds and their role in disease risk reduction-a review. Journal of Traditional and Complementary Medicine. 2018;**8**(4):451-458. DOI: 10.1016/j.jtcme.2017.08.006

[147] Mozaffarian D, Wu JHY. Flavonoids, dairy foods, and cardiovascular and metabolic health. A review of emerging biologic pathways. Circulation Research. 2018;**122**: 369-384. DOI: 10.1161/CIRCRESAHA.117.309008

[148] Khan A, Ikram M, Hahm JR, Kim M. Antioxidant and anti-inflammatory effects of *citrus* flavonoid Hesperetin: Special focus on neurological disorders. Antioxidants. 2020;**9**(7):609. DOI: 10.3390/antiox9070609

[149] Williams-Medina A, Deblock M, Janigro D. In vitro models of the blood–brain barrier: Tools in translational medicine. Frontiers in Medical Technology. 2021;**2**:623950. DOI: 10.3389/fmedt.2020.623950

[150] Pinent M, Castell-Auví A, Baiges I, Montagut Pino G, Arola L, Ardèvol A. Bioactivity of flavonoids on insulin-secreting cells. Comprehensive Reviews in Food Science and Food Safety. 2008;**7**: 299-308. DOI: 10.1111/j.1541-4337.2008.00048.x

[151] Kuryłowicz A. The role of Isoflavones in type 2 diabetes prevention and treatment-a narrative review. International Journal of Molecular Sciences. 2022;**22**(1): 218. DOI: 10.3390/ijms22010218

[152] Bondonno NP, Dalgaard F, Murray K, Davey RJ, Bondonno CP, Cassidy A, et al. Higher habitual flavonoid intakes are associated with a lower incidence of diabetes. The Journal of Nutrition. 2021;**151**(11):3533-3542. DOI: 10.1093/jn/nxab269

[153] Barańska A, Błaszczuk A, Polz-Dacewicz M, Kanadys W, Malm M, Janiszewska M, et al. Effects of soy Isoflavones on glycemic control and lipid profile in patients with type 2 diabetes: A systematic review and meta-analysis of randomized controlled trials. Nutrients. 2021;**13**(6):1886. DOI: 10.3390/nu13061886

[154] Oyenihi O, Nicole L, Oguntibeju O. Oxidative Stress and Diabetic Complications: The Role of Antioxidant

Vitamins and Flavonoids. 2014;923-931. DOI: 10.5772/57282

[155] Ribeiro D, Freitas M, Lima J, Fernandes E. Proinflammatory pathways: The modulation by flavonoids. Medicinal research reviews. 2015;**35**(5):877-936. DOI: 10.1002/med.21347

[156] Pallauf K, Duckstein N, Hasler M, Klotz LO, Rimbach G. Flavonoids as putative inducers of the transcription factors Nrf2, FoxO, and PPARγ. Oxidative Medicine and Cellular Longevity. 2017;**2017**:4397340. DOI: 10.1155/2017/4397340. Epub 2017 Jul 6

[157] Pournourmohammadi S, Grimaldi M, Stridh M, Lavallard V, Waagepetersen H, Wollheim C, et al. Epigallocatechin-3-gallate (EGCG) activates AMPK through the inhibition of glutamate dehydrogenase in muscle and pancreatic ß-cells: A potential beneficial effect in the pre-diabetic state? The International Journal of Biochemistry & Cell Biology. 2017;**88**: 220-225. DOI: 10.1016/j.biocel.2017.01.012

[158] Sok Yen F, Shu Qin C, Tan Shi Xuan S, Jia Ying P, Yi Le H, Darmarajan T, et al. Hypoglycemic effects of plant flavonoids: A review. Evidence-based Complementary and Alternative Medicine. 2021;**2021**: 2057333. DOI: 10.1155/2021/2057333

[159] Viollet B, Guigas B, Leclerc J, Hébrard S, Lantier Goëlzer L, Mounier R, et al. AMP-activated protein kinase in the regulation of hepatic energy metabolism: From physiology to therapeutic perspectives. Acta physiologica (Oxford, England). 2009; **196**:81-98. DOI: 10.1111/j.1748-1716.2009.01970.x

[160] Yang L, Wang Z, Jiang L, Sun W, Fan Q, Liu T. Total flavonoids extracted from *Oxytropis falcata* Bunge improve insulin resistance through regulation on the IKKβ/NF-κB inflammatory pathway. Evidence-based Complementary and Alternative Medicine. 2017;**2017**: 2405124. DOI: 10.1155/2017/2405124. Epub 2017 Mar 26

[161] Yang L, Wang Z, Jiang L, Sun W, Fan Q, Liu T. Total flavonoids extracted from Oxytropis falcata Bunge improve insulin resistance through regulation on the IKK ?/NF-? B inflammatory pathway. Evidence-Based Complementary and Alternative Medicine. 2017b;**1-6**. DOI: 10.1155/2017/2405124

[162] Kerimi A, Gauer JS, Crabbe S, Cheah JW, Lau J, Walsh R, et al. Effect of the flavonoid hesperidin on glucose and fructose transport, sucrase activity and glycaemic response to orange juice in a crossover trial on healthy volunteers. British Journal of Nutrition. 2019;**121**: 782-792

[163] Kapoor R, Kakkar P. Protective role of morin, a flavonoid, against high glucose induced oxidative stress mediated apoptosis in primary rat hepatocytes. PLoS One. 2012;**7**(8): e41663. DOI: 10.1371/journal.pone.0041663. Epub 2012 Aug 10

[164] Vinayagam R, Xu B. Antidiabetic properties of dietary flavonoids: A cellular mechanism review. Nutrition & Metabolism (London). 2015;**12**:60. DOI: 10.1186/s12986-015-0057-7

[165] Reis MB, Elias-Oliveira J, Pastore MR, Ramos SG, Gardinassi LG, Faccioli LH. Interleukin-1 receptor-induced nitric oxide production in the pancreas controls hyperglycemia caused by scorpion envenomation. Toxins (Basel). 2020;**12**(3):163. DOI: 10.3390/toxins12030163

[166] Heeba GH, Rabie EM, Abuzeid MM, Bekhit AA, Khalifa MM. Morin alleviates fructose-induced metabolic syndrome in rats *via* ameliorating oxidative stress, inflammatory and fibrotic markers. Korean Journal of Physiology Pharmacology. 2021;**25**(3):177-187. DOI: 10.4196/kjpp.2021.25.3.177

[167] Lamer-Zarawska E, Leszek J, Parvathaneni K, Yendluri B, Błach-Olszewska Z, Aliev G. Flavones from root of Scutellaria Baicalensis Georgi: Drugs of the future in neurodegeneration? CNS & Neurological Disorders Drug Targets. 2011;**10**:184-191. DOI: 10.2174/187152711794480384

[168] Wilcox G. Insulin and insulin resistance. Clinical Biochemist Reviews. 2005;**26**(2):19-39

[169] Vergun O, Svydenko LV, Grygorieva O, Shymanska O, Rakhmetov D, Brindza J, et al. Antioxidant capacity of plant raw material of Scutellaria baicalensis Georgi. Potravinarstvo Slovak Journal of Food Sciences. 2019;**13**(1)

[170] Kubatka P, Mazurakova A, Samec M, Koklesova L, Zhai K, AL-Ishaq R, et al. Flavonoids against non-physiologic inflammation attributed to cancer initiation, development, and progression—3PM pathways. EPMA Journal. 2021;**12**:559-587. DOI: 10.1007/s13167-021-00257-y

[171] Casanova E, Salvadó J, Crescenti A, Gibert-Ramos A. Epigallocatechin Gallate modulates muscle homeostasis in type 2 diabetes and obesity by targeting energetic and redox pathways: A narrative review. International Journal of Molecular Sciences. 2019;**20**(3):532. DOI: 10.3390/ijms20030532

[172] Daryabor G, Atashzar MR, Kabelitz D, Meri S, Kalantar K. The effects of type 2 diabetes mellitus on organ metabolism and the immune system. Frontiers in Immunology. 2020; **11**:1582. DOI: 10.3389/fimmu.2020.01582

[173] Casado-Díaz A, Rodríguez-Ramos Á, Torrecillas-Baena B, Dorado G, Quesada-Gómez JM, Gálvez-Moreno MÁ. Flavonoid Phloretin inhibits Adipogenesis and increases OPG expression in adipocytes derived from human bone-marrow mesenchymal stromal-cells. Nutrients. 2021;**13**(11): 4185. DOI: 10.3390/nu13114185

[174] Ansari P, Choudhury ST, Seidel V, Rahman AB, Aziz MA, Richi AE, et al. Therapeutic potential of quercetin in the Management of Type-2 diabetes mellitus. Life (Basel). 2022;**12**(8):1146. DOI: 10.3390/life12081146

[175] Herat LY, Matthews J, Azzam O, Schlaich MP, Matthews VB. Targeting features of the metabolic syndrome through sympatholytic effects of SGLT2 inhibition. Current Hypertension Reports. 2022;**24**(3):67-74. DOI: 10.1007/s11906-022-01170-z. Epub 2022 Mar 2

[176] Alkhalidy H, Moore W, Wang Y, Luo J, McMillan RP, Zhen W, et al. The flavonoid Kaempferol ameliorates Streptozotocin-induced diabetes by suppressing hepatic glucose production. Molecules. 2018;**23**(9):2338. DOI: 10.3390/molecules23092338

[177] Salehi B, Machin L, Monzote L, Sharifi-Rad J, Ezzat SM, Salem MA, et al. Therapeutic potential of quercetin: New insights and perspectives for human health. ACS Omega. 2020;**5**(20): 11849-11872. DOI: 10.1021/acsomega.0c01818

[178] Alaaeldin R, Abdel-Rahman IAM, Hassan HA, Youssef N, Allam AE, Abdelwahab SF, et al. Carpachromene

ameliorates insulin resistance in HepG2 cells via modulating IR/IRS1/PI3k/Akt/GSK3/FoxO1 pathway. Molecules. 2021; **26**(24):7629. DOI: 10.3390/molecules26247629

[179] Lee K, Chan JY, Liang C, Ip CK, Shi YC, Herzog H, et al. XBP1 maintains beta cell identity, represses beta-to-alpha cell transdifferentiation and protects against diabetic beta cell failure during metabolic stress in mice. Diabetologia. 2022;**65**(6):984-996. DOI: 10.1007/s00125-022-05669-7. Epub 2022 Mar 22

[180] Madhavan A, Kok BP, Rius B, Grandjean JMD, Alabi A, Albert V, et al. Pharmacologic IRE1/XBP1s activation promotes systemic adaptive remodeling in obesity. Nature Communications. 2022;**13**:608. DOI: 10.1038/s41467-022-28271-2

[181] Huang S, Xing Y, Liu Y. Emerging roles for the ER stress sensor IRE1α in metabolic regulation and disease. The Journal of Biological Chemistry. 2019; **294**(49):18726-18741. DOI: 10.1074/jbc.REV119.007036. Epub 2019 Oct 30

[182] Olefsky J. Treatment of insulin resistance with peroxisome proliferator-activated receptor ?? Agonists. The Journal of clinical investigation. 2000; **106**:467-472. DOI: 10.1172/JCI10843

[183] Faddladdeen KA, Ojaimi AA. Protective effect of pomegranate (*Punica granatum*) extract against diabetic changes in adult male rat liver: Histological study. Journnal of Microscopy Ultrastructure. 2019;7(4): 165-170. DOI: 10.4103/JMAU.JMAU_6_19. Epub 2019 Nov 18

[184] Riaz T, Junjappa R, Handigund M, Ferdous J, Kim HR, Chae H-J. Role of endoplasmic reticulum stress sensor IRE1α in cellular physiology, calcium, ROS signaling, and Metaflammation. Cell. 2020;**9**:1160. DOI: 10.3390/cells9051160

[185] Shah AA, Gupta A. Antioxidants in health and disease with their capability to defend pathogens that attack apple species of Kashmir. In: Ekiert HM, Ramawat KG, Arora J, editors. Plant Antioxidants and Health. Reference Series in Phytochemistry. Cham: Springer; 2021. DOI: 10.1007/978-3-030-45299-5_13-1

[186] Ponnampalam EN, Vahedi V, Giri K, Lewandowski P, Jacobs JL, Dunshea FR. Muscle antioxidant enzymes activity and gene expression are altered by diet-induced increase in muscle essential fatty acid (α-linolenic acid) concentration in sheep used as a model. Nutrients. 2019;**11**(4):723. DOI: 10.3390/nu11040723

[187] Cao Z-H, Wu Z, Hu C, Zhang M, Wang W-Z, Hu X-B. Endoplasmic reticulum stress and destruction of pancreatic β cells in type 1 diabetes. Chinese Medical Journal. 2020;**133**(1): 68-73. DOI: 10.1097/CM9.0000000000000583

[188] Khanna M, Agrawal N, Chandra R, Dhawan G. Targeting unfolded protein response: A new horizon for disease control. Expert Reviews in Molecular Medicine. 2021;**23**:E1. DOI: 10.1017/erm.2021.2

[189] Gupta A, Gupta R, Singh RL. Microbes and environment. In: Singh R, editor. Principles and Applications of Environmental Biotechnology for a Sustainable Future. Applied Environmental Science and Engineering for a Sustainable Future. Singapore: Springer; 2017. DOI: 10.1007/978-981-10-1866-4_3

[190] Russo M, Moccia S, Spagnuolo C, Tedesco I, Russo GL. Roles of flavonoids

against coronavirus infection. Chemico-Biological Interactions. 2020;**328**:109211. DOI: 10.1016/j.cbi.2020.109211. Epub 2020 Jul 28

[191] Wang M, Firrman J, Liu L, Yam K. A review on flavonoid Apigenin: Dietary intake, ADME, antimicrobial effects, and interactions with human gut microbiota. BioMed Research International. 2019; **2019**:7010467. DOI: 10.1155/2019/7010467

[192] Zakaryan H, Arabyan E, Oo A, Zandi K. Flavonoids: Promising natural compounds against viral infections. Archives of Virology. 2017;**162**(9): 2539-2551. DOI: 10.1007/s00705-017-3417-y. Epub 2017 May 25

[193] Musarra-Pizzo M, Pennisi R, Ben-Amor I, Mandalari G, Sciortino MT. Antiviral activity exerted by natural products against human viruses. Viruses. 2021;**13**:828. DOI: 10.3390/v13050828

[194] Hong J, Chi X, Yuan X, Wen F, Rai KR, Wu L, et al. I226R protein of African swine fever virus is a suppressor of innate antiviral responses. Viruses. 2022;**14**(3):575. DOI: 10.3390/v14030575

[195] Shibata C, Ohno M, Otsuka M, Kishikawa T, Goto K, Muroyama R, et al. The flavonoid apigenin inhibits hepatitis C virus replication by decreasing mature microRNA122 levels. Virology. 2014; **462–463**:42-48. DOI: 10.1016/j.virol.2014.05.024

[196] Evers D, Chao C-F, Wang X, Zhang Z, Huong S-M, Huang E-S. Human cytomegalovirus-inhibitory flavonoids: Studies on antiviral activity and mechanism of action. Antiviral Research. 2006;**68**:124-134. DOI: 10.1016/j.antiviral.2005.08.002

[197] Geng P, Zhu H, Zhou W, Su C, Chen M, Huang C, et al. Baicalin inhibits influenza a virus infection *via* promotion of M1 macrophage polarization. Frontiers in Pharmacology. 2020;**11**: 01298. DOI: 10.3389/fphar.2020.01298

[198] Huang L, Peng B, Nayak Y, Wang C, Si F, Liu X, et al. Baicalein and Baicalin promote melanoma apoptosis and senescence via metabolic inhibition. Frontiers in Cell and Development Biology. 2020;**8**:836. DOI: 10.3389/fcell.2020.00836 Erratum in: Front Cell Dev Biol.;10:876000

[199] Song J, Zhang L, Xu Y, Yang D, Zhang L, Yang S, et al. The comprehensive study on the therapeutic effects of baicalein for the treatment of COVID-19 in vivo and in vitro. Biochemical Pharmacology. 2021;**183**: 114302. DOI: 10.1016/j.bcp.2020.114302. Epub 2020 Oct 27

[200] Low ZX, OuYong BM, Hassandarvish P, Poh CL, Ramanathan B. Antiviral activity of silymarin and baicalein against dengue virus. Scientific Reports. 2021;**11**(1): 21221. DOI: 10.1038/s41598-021-98949-y

[201] Murtuja S, Shilkar D, Sarkar B, Sinha BN, Jayaprakash V. A short survey of dengue protease inhibitor development in the past 6 years (2015-2020) with an emphasis on similarities between DENV and SARS-CoV-2 proteases. Bioorganic & Medicinal Chemistry. 2021;**49**:116415. DOI: 10.1016/j.bmc.2021.116415. Epub 2021 Sep 20

[202] Tambunan BA, Priyanto H, Nugraha J. Soedarsono. CD4+ AND CD8 + T-cells expressing interfer Gama in active pulmonary tuberculosis patients. African Journal of Infectious Diseases. 2018;**12**(Suppl. 1):49-53. DOI: 10.2101/Ajid.12v1S.6

[203] Oo A, Rausalu K, Merits A, Higgs S, Vanlandingham D, Bakar SA, et al.

Deciphering the potential of baicalin as an antiviral agent for chikungunya virus infection. Antiviral Research. 2018; **150**:101-111. DOI: 10.1016/j.antiviral.2017.12.012. Epub 2017 Dec 19

[204] Jin H, Li D, Lin MH, Li L, Harrich D. Tat-based therapies as an adjuvant for an HIV-1 functional cure. Viruses. 2020;**12**(4):415. DOI: 10.3390/v12040415

[205] Qin S, Huang X, Qu S. Baicalin induces a potent innate immune response to inhibit respiratory syncytial virus replication via regulating viral non-structural 1 and matrix RNA. Frontiers in Immunology. 2022;**13**:907047. DOI: 10.3389/fimmu.2022.907047

[206] Wu C-C, Fang C-Y, Cheng Y-J, Hsu H-Y, Chou S-P, Huang S-Y, et al. Inhibition of Epstein-Barr virus reactivation by the flavonoid apigenin. Journal of Biomedical Science. 2017;**24**(1):1-13. DOI: 10.1186/s12929-016-0313-9

[207] Wang S, Ling Y, Yao Y, et al. Luteolin inhibits respiratory syncytial virus replication by regulating the MiR-155/SOCS1/STAT1 signaling pathway. Virology Journal. 2020;**17**:187. DOI: 10.1186/s12985-020-01451-6

[208] Silva Dos Santos J, Gonçalves Cirino JP, de Oliveira CP, Ortega MM. The pharmacological action of Kaempferol in central nervous system diseases: A review. Frontiers in Pharmacology. 2021;**11**:565700. DOI: 10.3389/fphar.2020.565700

[209] Di Petrillo A, Orrù G, Fais A, Fantini MC. Quercetin and its derivates as antiviral potentials: A comprehensive review. Phytotherapy Research. 2021; **36**(1):266-278. DOI: 10.1002/ptr.7309. Epub 2021 Oct 28

[210] Rawangkan A, Kengkla K, Kanchanasurakit S, Duangjai A, SaoII99kaew S. Anti-influenza with green tea Catechins: A systematic review and meta-analysis. Molecules. 2021; **26**(13):4014. DOI: 10.3390/molecules 26134014

[211] Farazuddin M, Mishra R, Jing Y, Srivastava V, Comstock AT, Sajjan US. Quercetin prevents rhinovirus-induced progression of lung disease in mice with COPD phenotype. PLoS One. 2018;**13**(7): e0199612. DOI: 10.1371/journal.pone.0199612

[212] Jung K, Saif LJ, Wang Q. Porcine epidemic diarrhea virus (PEDV): An update on etiology, transmission, pathogenesis, and prevention and control. Virus Research. 2020; **2020**(286):198045. DOI: 10.1016/j.virusres.;198045. Epub 2020 Jun 2

[213] Anand AV, Balamuralikrishnan B, Kaviya M, Bharathi K, Parithathvi A, Arun M, et al. Medicinal plants, phytochemicals, and herbs to combat viral pathogens including SARS-CoV-2. Molecules. 2021;**26**:1775. DOI: 10.3390/molecules 26061775

[214] Kim C-H. Anti–SARS-CoV-2 natural products as potentially therapeutic agents. Frontiers in Pharmacology. 2021;**12**:590509. DOI: 10.3389/fphar.2021.590509

[215] Khazeei Tabari MA, Iranpanah A, Bahramsoltani R, Rahimi R. Flavonoids as promising antiviral agents against SARS-CoV-2 infection: A mechanistic review. Molecules. 2021;**26**:3900. DOI: 10.3390/molecules26133900

[216] Vanderlinden E, Naesens L. Emerging antiviral strategies to interfere with influenza virus entry. Medicinal Research Reviews. 2014;**34**(2):301-334. DOI: 10.1002/med.21289

[217] Kaihatsu K, Yamabe M, Ebara Y. Antiviral mechanism of action of

Epigallocatechin-3-*O*-gallate and its fatty acid esters. Molecules. 2018;**23**(10): 2475. DOI: 10.3390/molecules23102475

[218] Song JM. Anti-infective potential of catechins and their derivatives against viral hepatitis. Clinical and Experimental Vaccine Research. 2018;7(1):37-42. DOI: 10.7774/cevr.2018.7.1.37

[219] Nabil-Adam A, Shreadah M. Biogenic silver nanoparticles synthesis from new record aquatic bacteria of Nile tilapia and evaluation of their biological activity. Journal of Pure and Applied Microbiology. 2020;**14**:2491-2511. DOI: 10.22207/JPAM.14.4.27

[220] Wang YQ, Li QS, Zheng XQ, Lu JL, Liang YR. Antiviral effects of green tea EGCG and its potential application against COVID-19. Molecules. 2021; **26**(13):3962. DOI: 10.3390/ molecules26133962

[221] Khachatoorian R, Arumugaswami V, Raychaudhuri S, Yeh G, Maloney E, Wang J, et al. Divergent antiviral effects of bioflavonoids on the hepatitis C virus life cycle. Virology. 2012;**433**:346-355. DOI: 10.1016/j.virol.2012.08.029

[222] Liu S, Li H, Tang M, Cao Y. (−)-Epigallocatechin-3-gallate inhibition of Epstein-Barr virus spontaneous lytic infection involves downregulation of latent membrane protein 1. Experimental and Therapeutic Medicine. 2018;**15**(1):1105-1112. DOI: 10.3892/ etm.2017.5495. Epub 2017 Nov 13

[223] Baz M, Boivin G. Antiviral agents in development for Zika virus infections. Pharmaceuticals. 2019;**12**:101. DOI: 10.3390/ph12030101 www.mdpi. com/journal/pharmaceuticals

[224] Jannat K, Paul AK, Bondhon TA, Hasan A, Nawaz M, Jahan R, et al. Nanotechnology applications of flavonoids for viral diseases. Pharmaceutics. 2021;**13**(11):1895. DOI: 10.3390/pharmaceutics13111895

[225] Arabyan E, Hakobyan A, Kotsinyan A, Karalyan Z, Arakelov V, Arakelov G, et al. Genistein inhibits African swine fever virus replication in vitro by disrupting viral DNA synthesis. Antiviral Research. 2018;**156**: 128-137. DOI: 10.1016/j.antiviral.2018. 06.014. Epub 2018 Jun 22

Chapter 4

Purple Corn Cob: Rich Source of Anthocyanins with Potential Application in the Food Industry

Andreea Stănilă, Teodora Daria Pop and Zorița Maria Diaconeasa

Abstract

As every year, the entire food chain generates huge amounts of food loss and waste, and there is a great interest in solving the inefficient waste management by implementing the sustainability concept for achieving "waste-to-wealth" goal. This refers to recovering renewable bioactive compounds from food wastes in order to use them as low-cost source of value-added ingredients for different industries. In this way, this work focuses its attention on purple corn cob, a by-product that was not very used in food industry. Purple corn has gained attention due to its capability of coloring food and beverages and the evidence of the antioxidant, anti-inflammatory, and cardiovascular health benefits. As the production is growing year by year, the amounts of waste produced is rising. As a result, purple corn cob caught our attention, reason why in this study we concentrate to summarize and emphasize the compounds that give the color of this waste, anthocyanins.

Keywords: anthocyanins, bioactive compounds, food waste, purple corn cob, food industry

1. Introduction

To date, adding coloring to food and drink color is as significant as it has always been, especially in a society where the quantity of processed food has increased rapidly over the previous half-century. Being the first attribute that potential buyers are aware of, color plays a vital role in motivating the customer to purchase the product [1]. Over the last few years, the food industry's interest in natural colorants replacing synthetic ones has increased, mainly due to safety issues [2]. Among natural alternatives for synthetic dyes, anthocyanins represent an important class of compounds that provide red to a blue color to food by incorporating into aqueous systems. Also, in the European Union and Japan, anthocyanins are recognized as food colorants with the code E-163 [3].

Purple corn (*Zea mays* L.) represents a rich source of anthocyanins, originated in South America, mainly Peru, and cultivated also in Ecuador, Bolivia, and Argentina. Generally, it is used for the preparation of traditional drinks and desserts, generating important quantities of residues, and its disposal involves major economic expenses [4, 5]. Besides this, generated wastes can cause major environmental problems due to their high organic charge. However, over the last decades, using these residues has been encouraged as they are good sources of potentially useful bioactive chemicals, and valorizing them might be a viable technique for mitigating their environmental implications and thereby improving the food industry's sustainability [5].

As a major route for their valorization, the recovery of bioactive molecules from food by-products has gained popularity. In this framework, the present review aims to highlight purple cob's capability to be an available source of bioactive compounds with several potential applications, focusing our work on its food industry utilization.

2. Purple corn cob is a rich source of bioactive compounds

The current state of knowledge shows us that purple corn cob represents a rich source of phenolic compounds, especially phenolic acids, flavonoids, and anthocyanins [6, 7].

Purple corn gained attention mainly because of its rich content in anthocyanins, but comparing the seeds (0.5–6.8 mg/g fresh weight) with the cob, the reported values regarding anthocyanins content were higher for the latest (0.8–71.5 mg/g fresh weight) [4].

Based on the previous statement, our focus is on a purple corn cob, as the anthocyanins content makes them potential contributors as natural colorants, varying from blue to red tonalities. Despite their great coloring capacity, these compounds have been increasingly used in the food industry due to their ability to confer bioactive properties to the products [8]. Along with a great coloring capacity, anthocyanins act as antioxidants and antibacterial compounds and help prevent cardiovascular diseases, cancer, diabetes, and have neuroprotective effects [9–13].

Regarding the anthocyanins content, some researchers characterized the purple corn cob. Among them, Díaz-García et al., obtained higher values than other studies, 41.32 ± 0.95 mg C3GE/g DW, compared with 9.30–15.16 mg C3GE/g DW, both using pH differential measurement [4, 14]. Using the same method, Lao and Giusti [15] obtained for the purple corn cob values ranging from 3.1 to 100.3 mg C3G/g. By using a conventional spectrophotometric method, Pascual-Teresa et al. obtained 34 g per 100 g powder, expressed in cyanidin-3-monoglucoside [16]. Another study showed that the anthocyanin content of purple corn cob was calculated to be 92.3 ± 2.1 mg/100 g, expressed in cyanidin-3-glucoside [17]. A much higher value of anthocyanins was presented in another study, where the authors obtained 1219.4 mg/100 g [18]. The comparison between the anthocyanin contents in purple corn cob is difficult as there are several factors to consider, including the harvesting region, methods used by researchers, storage time, and genetic differences [19].

The profile of anthocyanins present in purple corn cob has been characterized by Fernandez-Aulis et al., by identifying the structures using HPLC method, are fragments corresponding to cyanidin, pelargonidin, and peonidin, and they are: cyanidin-3-glucoside, cyanidin-3-(6″-malonyl) glucoside, pelargonidin-3-glucoside, peonidin-3-glucoside, pelargonidin-3-(6″-malonyl) glucoside, and peonidin-3-(6″-malonyl) glucoside, which was in accordance with the previous reports. Among these, the main anthocyanin was cyanidin-3-(6″-malonyl) glucoside obtained in case of enzymatic assisted extraction [20]. Moreover, cyanidin-3-(6″-ethylmalonyl) glucoside, pelargonidin-3-(6″-ethylmalonyl) glucoside, and peonidin-3-(6″-ethylmalonyl) glucoside are three more compounds identified by Pascual-Teresa et al. besides those mentioned above [16].

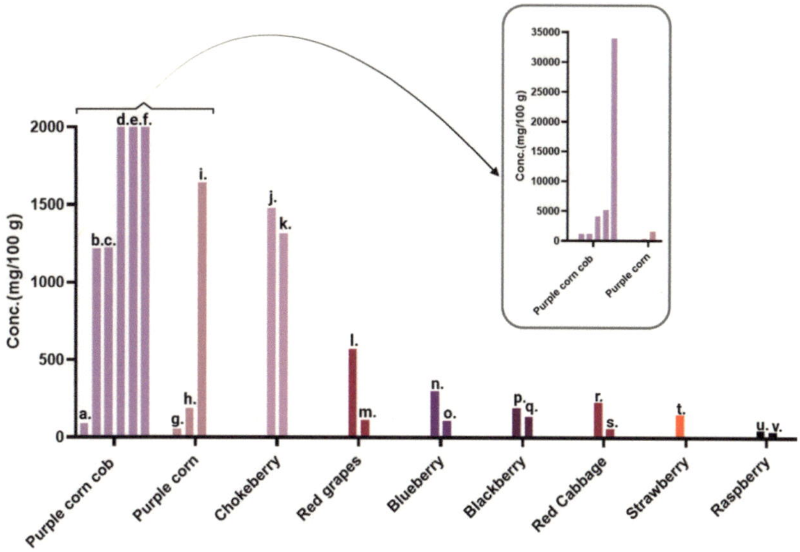

Figure 1.
Comparison between rich in anthocyanin sources and purple corn cob based on the concentration (mg/100 g), reported by different authors a—[17], b—[18], c—[14], d—[4], e—[15], f—[16], g—[17], h—[7], i—[21], j—[22], k—[23], l—[24], m—[25], n—[26], o—[27], p—[28], q—[29], r—[30], s—[31], t—[32], u—[33], v—[34].

The reported results suggest that purple corn cob can be considered a promising source of anthocyanins, and this statement can be strengthened by comparing values with other dietary sources, remarked for their rich content in anthocyanins. As shown in **Figure 1**, purple corn cob is a valuable waste that can be successfully used as a rich source of anthocyanins, and by this, an important number of applications can be taken into consideration.

Throughout the years, purple corn gained attention due to its recognized antioxidant property, becoming consumed as food, but also incorporated in new products, fact highlighted too by international data bases, as in 2009, Peru exported total value of US$9,782,564, while in 2013, the export reached the approximate value of US$17,981,398 [19, 35]. In 2015, AgriFutures Australia reported a total of 7000 tones as being traded globally each year, with the majority producer, Peru, and for 348 tones, China [36]. National Institute of Statistics and Information Science analyzed the purple corn production index from 1990 to 2018, highlighting the fact that the data reached an all-time high in April 2018 [37]. As the production is rising and the consumption per capita has increased from 6.8 to 12.3 kg per year [19], the amount of generated waste is also recording higher values, the cob representing 15% of the purple corn ear [38]. Also, taking into consideration its richness in bioactive compounds and its growing range of applications, purple corn cob is becoming a topic of interest for future studies.

3. Anthocyanin recovered from by-products

3.1 Extraction methods

As anthocyanins are found as secondary plant metabolites in the case of purple corn cob, it is important to take into consideration their extraction and isolation from the tissue with adequate method, as their main issue is their high susceptibility to degradation, when isolated. Among the factors that affect anthocyanins' stability, the most referred are pH, when it has values above 7, explaining the requirement of acidified environment for most of the extraction protocols and temperature. The last one impacts the levels of anthocyanin extracted, a higher temperature being likely to induce degradation and impact in this way the extraction factor [39]. An ideal extraction procedure for food-grade anthocyanins would maximize pigment output while reducing degradation and changing the natural state of the target compounds [40]. When choosing the extraction method, it is important to take into consideration the application, as for food industry the solvents play a decisive role. In most of the times, anthocyanins' extraction is conducted by grinding, drying or lyophilizing the matrices, or soaking fresh materials into solvents such as water, ethanol, methanol, acetone, or others. As mentioned, the solvents are frequently acidified to make the extraction process easier and to keep the pigments stable during the process. Other possible influences on anthocyanin extraction that should be considered beside pH and temperature are time, solid:liquid ratio, assistance with microwave, ultrasound, and sonication [3]. In the case of purple corn cob, different optimized extraction techniques have been developed, both conventional and new methods (**Table 1**), taking into consideration the time, temperature, and solid: liquid ratio. In this way, Yang and Zhai [17] macerated the matrices and used methanol as solvent, the same one used by Li et al. in their study [41]. Another solvent used by Nisi et al. is acetone, but due to safety concerns, these two, methanol and acetone, are not desirable if the extract is used in food applications. The same author tested also the efficacy of water and ethanol solvent, by comparing with acetone, and the results were similar [42]. Also, Lao and Giusti optimized an extraction protocol using a combination of water and ethanol (EtOH), which is food-friendly. Besides analyzing the best extraction ratio, they focused also on its acidity in order to increase extraction efficiency [3]. Yang et al.'s work focused on choosing the optimum conditions for the extraction of anthocyanins from purple corn cob taking into account the main factors related to the process, including solvent, acid, solvent concentration, and acid concentration, using a mixture of solvent and water. The maximum yield for purple corn cob extraction was obtained with 80% methanol and 1% citric acid, and using ethanol, the best conditions were obtained with 80% solvent and 0.5% citric acid [43]. Based on safety concerns related to certain organic solvents, Rajha et al. managed to study the efficiency in terms of anthocyanins' extraction using β-cyclodextrin-water as solvent, β-cyclodextrin being generally recognized as safe (GRAS) solvent. They suggested a low-cost extraction protocol for the valorization of purple corn cobs, obtaining the best extraction rate using 39.8 mg/mL β-cyclodextrin at 68.8°C for 60.4 min [44].

Recently, there is a focus on reducing or eliminating the use of toxic solvents by taking advantage of the potential that emerging extraction strategies have, which are capable of reducing the processing time, maintaining also the bioactivity of the compounds [48]. Piyapanrungrueang et al. realized a comparison study between the conventional extraction methods and the emerging ones, having a purple corn cob hybrid as matrices. They characterized and optimized methods for conventional, ultrasound-assisted extraction, microwave-assisted one, and for ohmic heating. By taking into consideration the amount of anthocyanin extracted, energy efficiency, and color value, the best method for extracting anthocyanin from purple corn cob is the microwave-assisted one [45]. There are also other authors that optimized emerging methods for extracting anthocyanins as presented in **Table 1**. When choosing the extracting protocol, the future applications should be taken into consideration, as

Method	Solvent	Acidity	Time	Temp (°C)	Solid:liquid ratio	Anthocyanin content mg/100 g (DW)	Ref.
Conventional extraction methods	Methanol (MeOH)	0%	24h	4°C	1:4	92.3 ± 2.1	[17]
	MeOH	1% HCl	12h/extraction (3 times)	4°C	1:10	0.49–4.6%	[41]
	Acetone 70%	0.01% HCl	12 h	RT	1:50 w/v	931 ± 44	[42]
	EtOH 50%	0.01% HCl 6M	Overnight	RT	1:50 w/v	854 ± 24	[42]
	Water	0.01% HCl	90 min	RT	1:20	1004.22 ± 10.3	[18]
					1:50	754.15 ± 4.5	
					1:100	735.58 ± 11.35	
	Water	0.01% HCl	Not mentioned	RT	—	610	[3]
	EtOH	0.01% HCl	Not mentioned	RT	—	340	
	20% EtOH	0.01% HCl	Not mentioned	RT	—	1060	
	50% EtOH	0.01% HCl	Not mentioned	RT	—	1430	
	80% EtOH	0.01% HCl	Not mentioned	RT	—	1190	
	80% EtOH	0.5% $C_6H_8O_7$	24 h	4°C	1:50	485	[43]
	90% EtOH	0.5% $C_6H_8O_7$	24h	4°C	1:50	262	
	80% MeOH	1% $C_6H_8O_7$	24h	4°C	1:50	590	
	80% EtOH	0.5 % CH_3COOH	24h	4°C	1:50	423	
	80% MeOH	0.5% CH_3COOH	24h	4°C	1:50	504	
	β-cyclodextrin (25 mg/mL)	0%	29.9 min	52.5°C	Not mentioned	1439	[44]
	β-cyclodextrin (25 mg/mL)	0%	105 min	79.9°C	Not mentioned	1557	
	β-cyclodextrin (39.8 mg/mL)	0%	149.6 min	36.2°C	Not mentioned	1404	
	β-cyclodextrin (39.8 mg/mL)	0%	60.4 min	68.8°C	Not mentioned	2608	
	β-cyclodextrin (49.8 mg/mL)	0%	105 min	52.5°C	Not mentioned	2497 mg/100g	
	Water	0%	5 min	70°C	1:20	366.1 ± 4.7	[45]

Method	Solvent	Acidity	Time	Temp (°C)	Solid:liquid ratio	Anthocyanin content mg/100 g (DW)	Ref.
Emerging extraction methods	Ultrasound-assisted extraction						
	50% EtOH	0%	30 min	65°C	1:20	240 mg/100 g	[46]
	Water	0%	5 min	80°C	1:20	382.7 ± 9.1	[45]
	80% EtOH	1% lactic acid	20 min	25°C	1:10	2431	[20]
	Microwave-assisted extraction						
	95% EtOH	1.5 M HCl	19 min	RT	1:20	185.1	[47]
	90% EtOH	1% lactic acid	1 min	25°C	1:10	1977	[20]
	Water	0%	2 min	83–91°C	1:20	397.1 ± 11.6	[45]
	Ohmic heating extraction						
	Water	0%	3.5 min	80°C (220V)	1:20	328.0 ± 4.5	[45]

Table 1.
Optimized extraction techniques: conventional and new methods.

there are plenty possibilities for anthocyanin from purple corn cob. Related to this, the next chapter focuses its attention on studies that emphasize purple corn cob's applications in different industries.

3.2 Recent studies and future perspective

In terms of using the cobs in order to minimize the waste production, Nisi et al. proposed a quick and cheap extraction method to obtain anthocyanin with the aim of using them as dye for different natural fibers. Moreover, they managed to recover and use all the material to produce nutraceuticals and pigmented litter for pets [42]. In this framework of the biorefinery approach in valorizing purple corn cob, authors proposed a natural dye for fabrics, showing also ultraviolet protection of the dyed clothes. They present the final product as having good durability, acceptable stability, excellent aesthetic appearance and also sustainable, since are dyed with natural pigments. Also, they demonstrated that the extract possesses good anti-inflammatory properties, highlighting the possibility of incorporating in food products to take advantage of its full potential related to health benefits. The purple lignocellulosic solid residue is considered feasible for animal bedding, which can be compostable, nulling in this way the waste produced [42].

Within the same frame, Gullón et al. characterized purple corn cob's potential in order to comprehend whether it can be used as multifunctional ingredient for food and pharmaceutical industries and what properties does it possess [5]. After the extraction conditions of bioactive compounds from purple corn cob were successfully performed, oligosaccharides and phenolic compounds, including flavonoids and anthocyanins were characterized. The bioactive compounds showed complex structures as the extract was stable at high temperatures when subjected to thermogravimetric analysis. Overall, their research concluded that purple corn cob represents a sustainable source of bioactive compounds with economical value, therefore improving the food industry's competitiveness [5].

Regarding its culinary applications, purple corn cob is used as a base ingredient in the process of obtaining traditional drink *chicha morada* and dessert named *mazamorra morada.* For both of then, the whole corn including the corn cob is boiled in order to extract the color. To obtain the drink, some fruits, spices, and sugar are mixed together. Besides these, the dessert requires a binder followed by cooking. Regarding the heat treatments that can affect the product nutritional quality, some studies concluded that thermal processing such as boiling, roasting, or frying is associated with a decrease in bioactive compounds, as they are not stable in such conditions [19].

Also related to food industry, a recent study developed and optimized a low-calorie tea formulation containing purple corn cob, stevia as sweetener, cinnamon, and clove as flavoring and quince as a pH regulator that improves the color. The product exhibits enhanced antioxidant capacity, which, in future studies, will be evaluated on human volunteers in order to offer information on the true impact of the new antioxidant product [4]. In a similar area of products, Wattanathorn et al. [49] obtained a functional drink containing the extract of purple corn cob and pandan leaves, being the first study that showed that a polyphenol-rich supplement can improve cognitive function in a menopausal animal model. The functional drink has a cognitive boosting effect that is comparable to donepezil, a common medicine used to treat memory problems today. In this way, the product based on purple corn cob extract and pandan leaves can be used as a viable supplement to lower the risk of memory deterioration in menopausal women, which is simple to achieve, based on its advantages and low raw material prices. Clinical trial research and subchronic toxicity, one the other hand, are necessary [49].

As presented, purple corn cob is a valuable source of bioactive compounds, with different possible applications, but related to food industry, several studies need to be made in order to benefits from its full potential.

4. Conclusions

As one of the humanity's greatest difficulties is to live in a society without hunger but with high quality and safe food, it can be assumed that global food loss and waste must be dramatically decreased, because the inefficient waste management causes environmental damages. In this framework, circular bioeconomy represents a great potential approach for reducing these insufficiencies by implementing the term "waste-to-wealth," which refers to the converting of renewable biological supplies with high-end commerce, into valuable resources for a longer period of time, with no waste generation. In this context, agri-food by-products and wastes are of great interest as they are rich in bioactive compounds, which gives them multiple applications.

Acknowledgements

This research was funded by the Romanian National Authority for Scientific Research (UEFISCDI), grant numbers PN-III-P1-1.1-TE-2019-0960/178TE/2020.

Conflict of interest

The authors declare no conflict of interest.

Acronyms and abbreviations

C3GE	cyanidin-3-glucoside
DW	dry weight
RT	room temperature

Author details

Andreea Stănilă, Teodora Daria Pop and Zorița Maria Diaconeasa*
University of Agricultural Sciences and Veterinary Medicine, Cluj-Napoca, Romania

*Address all correspondence to: zorita.sconta@usamvcluj.ro

References

[1] Stich E. 1 - Food color and coloring food: Quality, differentiation and regulatory requirements in the European Union and the United States. In: Carle R, Schweiggert RM, editors. Handbook on Natural Pigments in Food and Beverages. Woodhead Publishing; 2016. pp. 3-27

[2] Mahdavi SA, Jafari SM, Ghorbani M, Assadpoor E. Spray-drying microencapsulation of anthocyanins by natural biopolymers: A review. Drying Technology. 2014;**32**(5):509-518

[3] Lao F, Giusti MM. Extraction of purple corn (*Zea mays* L.) cob pigments and phenolic compounds using food-friendly solvents. Journal of Cereal Science. 2018;**80**:87-93

[4] Díaz-García A, Salvá-Ruíz B, Bautista-Cruz N, Condezo-Hoyos L. Optimization of a natural low-calorie antioxidant tea prepared from purple corn (*Zea mays* L.) cobs and stevia (Stevia rebaudiana Bert.). LWT. 2021;**150**:111952

[5] Gullón P, Eibes G, Lorenzo JM, Pérez-Rodríguez N, Lú-Chau TA, Gullón B. Green sustainable process to revalorize purple corn cobs within a biorefinery frame: Co-production of bioactive extracts. Science of the Total Environment. 2020;**709**:136236

[6] Ranilla LG, Rios-Gonzales BA, Ramírez-Pinto MF, Fuentealba C, Pedreschi R, Shetty K. Primary and phenolic metabolites analyses, in vitro health-relevant bioactivity and physical characteristics of purple corn (*Zea mays* L.) grown at two andean geographical locations. Metabolites. 2021;**11**(11)

[7] Ramos-Escudero F, Muñoz AM, Alvarado-Ortíz C, Alvarado Á, Yáñez JA. Purple corn (*Zea mays* L.) phenolic compounds profile and its assessment as an agent against oxidative stress in isolated mouse organs. Journal of Medicinal Food. 2012;**15**(2):206-215

[8] Vega EN, Molina AK, Pereira C, Dias MI, Heleno SA, Rodrigues P, et al. Anthocyanins from Rubus fruticosus L. and Morus nigra L. Applied as food colorants: A natural alternative. Plants. 2021;**10**(6):1181

[9] Diaconeasa Z, Leopold L, Rugină D, Ayvaz H, Socaciu C. Antiproliferative and antioxidant properties of anthocyanin rich extracts from blueberry and blackcurrant juice. International Journal of Molecular Sciences. 2015;**16**(2):2352-2365

[10] Kimble R, Keane KM, Lodge JK, Howatson G. Dietary intake of anthocyanins and risk of cardiovascular disease: A systematic review and meta-analysis of prospective cohort studies. Critical Reviews in Food Science and Nutrition. 2019;**59**(18):3032-3043

[11] Salehi B, Sharifi-Rad J, Cappellini F, Reiner Ž, Zorzan D, Imran M, et al. The therapeutic potential of anthocyanins: Current approaches based on their molecular mechanism of action. Frontiers in Pharmacology. 2020;**11**:1300

[12] Zhang J, Wu J, Liu F, Tong L, Chen Z, Chen J, et al. Neuroprotective effects of anthocyanins and its major component cyanidin-3-O-glucoside (C3G) in the central nervous system: An outlined review. European Journal of Pharmacology. 2019;**858**:172500

[13] Sun X-h, Zhou T-t, Wei C-h, Lan W-q, Zhao Y, Pan Y-j, et al. Antibacterial

effect and mechanism of anthocyanin rich Chinese wild blueberry extract on various foodborne pathogens. Food Control. 2018;**94**:155-161

[14] Jing P, Noriega V, Schwartz SJ, Giusti MM. Effects of growing conditions on purple corncob (*Zea mays* L.) anthocyanins. Journal of Agricultural and Food Chemistry. 2007;**55**(21):8625-8629

[15] Lao F, Giusti MM. Quantification of purple corn (*Zea mays* L.) anthocyanins using spectrophotometric and HPLC approaches: Method comparison and correlation. Food Analytical Methods. 2016;**9**(5):1367-1380

[16] de Pascual-Teresa S, Santos-Buelga C, Rivas-Gonzalo JC. LC–MS analysis of anthocyanins from purple corn cob. Journal of the Science of Food and Agriculture. 2002;**82**(9):1003-1006

[17] Yang Z, Zhai W. Identification and antioxidant activity of anthocyanins extracted from the seed and cob of purple corn (*Zea Mays* L.). Innovative Food Science and Emerging Technologies. 2010;**11**(1):169-176

[18] Vayupharp B, Laksanalamai V. Antioxidant properties and color stability of anthocyanin purified extracts from Thai Waxy purple corn cob. Journal of Food and Nutrition Research. 2015;**3**(10):629-636

[19] Salvador-Reyes R, Clerici MTPS. Peruvian Andean maize: General characteristics, nutritional properties, bioactive compounds, and culinary uses. Food Research International. 2020;**130**:108934

[20] Fernandez-Aulis F, Hernandez-Vazquez L, Aguilar-Osorio G, Arrieta-Baez D, Navarro-Ocana A. Extraction and identification of anthocyanins in corn cob and corn husk from cacahuacintle maize. Journal of Food Science. 2019;**84**(5):954-962

[21] Cevallos-Casals BA, Cisneros-Zevallos L. Stoichiometric and kinetic studies of phenolic antioxidants from Andean purple corn and red-fleshed sweetpotato. Journal of Agricultural and Food Chemistry. 2003;**51**(11):3313-3319

[22] Wu X, Gu L, Prior RL, McKay S. Characterization of anthocyanins and proanthocyanidins in some cultivars of Ribes, Aronia, and Sambucus and their antioxidant capacity. Journal of Agricultural and Food Chemistry. 2004;**52**(26):7846-7856

[23] Rugină D, Diaconeasa Z, Coman C, Bunea A, Socaciu C, Pintea A. Chokeberry anthocyanin extract as pancreatic β-cell protectors in two models of induced oxidative stress. Oxidative Medicine and Cellular Longevity. 2015;**2015**:429075

[24] Kallithraka S, Aliaj L, Makris DP, Kefalas P. Anthocyanin profiles of Major Red Grape (*Vitis vinifera* L.) varieties cultivated in greece and their relationship with in vitro antioxidant characteristics. International Journal of Food Science and Technology. 2009;**44**(12):2385-2393

[25] Muñoz-Espada AC, Wood KV, Bordelon B, Watkins BA. Anthocyanin quantification and radical scavenging capacity of concord, norton, and marechal foch grapes and wines. Journal of Agricultural and Food Chemistry. 2004;**52**(22):6779-6786

[26] Mustafa AM, Angeloni S, Abouelenein D, Acquaticci L, Xiao J, Sagratini G, et al. A new HPLC-MS/MS method for the simultaneous determination of 36 polyphenols in blueberry, strawberry and their commercial products and determination

of antioxidant activity. Food Chemistry. 2022;**367**:130743

[27] Yousef GG, Brown AF, Funakoshi Y, Mbeunkui F, Grace MH, Ballington JR, et al. Efficient quantification of the health-relevant anthocyanin and phenolic acid profiles in commercial cultivars and breeding selections of blueberries (Vaccinium Spp.). Journal of Agricultural and Food Chemistry. 2013;**61**(20):4806-4815

[28] Jara-Palacios MJ, Santisteban A, Gordillo B, Hernanz D, Heredia FJ, Escudero-Gilete ML. Comparative study of red berry pomaces (blueberry, red raspberry, red currant and blackberry) as source of antioxidants and pigments. European Food Research and Technology. 2019;**245**(1):1-9

[29] Fan-Chiang H-J, Wrolstad RE. Anthocyanin pigment composition of blackberries. Journal of Food Science. 2005;**70**(3):C198-C202

[30] Scalzo RL, Genna A, Branca F, Chedin M, Chassaigne H. Anthocyanin composition of cauliflower (*Brassica oleracea* L. var. botrytis) and cabbage (*B. oleracea* L. var. capitata) and its stability in relation to thermal treatments. Food Chemistry. 2008;**107**(1):136-144

[31] Wiczkowski W, Szawara-Nowak D, Topolska J. Red cabbage anthocyanins: Profile, isolation, identification, and antioxidant activity. Food Research International. 2013;**51**(1):303-309

[32] Dzhanfezova T, Barba-Espín G, Müller R, Joernsgaard B, Hegelund JN, Madsen B, et al. Anthocyanin profile, antioxidant activity and total phenolic content of a strawberry (Fragaria × ananassa Duch) genetic resource collection. Food Bioscience. 2020;**36**:100620

[33] Szymanowska U, Baraniak B, Bogucka-Kocka A. Antioxidant, anti-inflammatory, and postulated cytotoxic activity of phenolic and anthocyanin-rich fractions from Polana Raspberry (*Rubus idaeus* L.) fruit and juice—In Vitro study. Molecules. 2018;**23**(7):1812

[34] Pantelidis G, Vasilakakis MD, Manganaris GA, Diamantidis G. Antioxidant capacity, phenol, anthocyanin and ascorbic acid contents in raspberries, blackberries, red currants, gooseberries and Cornelian cherries. Food Chemistry. 2007;**102**:777-783

[35] UNEP. Peru's Sustainable Trade Potential: Biodiversity-Based Products. 2015

[36] O'hare T, Fanning K, Sanderson J, Naidoo R. Feasibility and Opportunities for Peruvian Purple Corn in Australia. 2015

[37] Production PA. Index: Cereals: Purple Corn | Economic Indicators | CEIC. Available from: https://www.ceicdata.com/en/peru/agricultural-production-index-2007100/agricultural-production-index-cereals-purple-corn. [Accessed: March 22, 2022]

[38] Cristianini M, Guillén Sánchez JS. Extraction of bioactive compounds from purple corn using emerging technologies: A review. Journal of Food Science. 2020;**85**(4):862-869

[39] Silva S, Costa EM, Calhau C, Morais RM, Pintado ME. Anthocyanin extraction from plant tissues: A review. Critical Reviews in Food Science and Nutrition. 2017;**57**(14):3072-3083

[40] Lao F, editor. Purple Corn (*Zea mays* L.) Cob Anthocyanins: Extraction, Quantification, Spray Drying and Complexation with Proteins. 2016

[41] Li CY, Kim HW, Won SR, Min HK, Park KJ, Park JY, et al. Corn husk as a

potential source of anthocyanins. Journal of Agricultural and Food Chemistry. 2008;**56**(23):11413-11416

[42] De Nisi P, Borlini G, Parizad PA, Scarafoni A, Sandroni P, Cassani E, et al. Biorefinery approach applied to the valorization of purple corn cobs. ACS Sustainable Chemistry & Engineering. 2021;**9**(10):3781-3791

[43] Yang Z, Fan G, Gu Z, Han Y, Chen Z. Optimization extraction of anthocyanins from purple corn (*Zea mays* L.) cob using tristimulus colorimetry. European Food Research and Technology. 2008;**227**(2):409-415

[44] Rajha HN, Khabbaz S, Rached RA, Debs E, Maroun RG, Louka N, editors. Optimization of polyphenols extraction from purple com cobs using ß-cyclodextrin as a green solvent. In: 2020 5th International Conference on Renewable Energies for Developing Countries (REDEC). 29-30 June 2020

[45] Piyapanrungrueang W, Chantrapornchai W, Haruthaithanasan V, Sukatta U, Aekatasanawan C. Comparison of anthocyanin extraction methods from high anthocyanin purple corn cob hybrid: KPSC 901, and quality of the extract powder. Journal of Food Processing and Preservation. 2016;**40**(5):1125-1133

[46] Muangrat R, Pongsirikul I, Blanco PH. Ultrasound assisted extraction of anthocyanins and total phenolic compounds from dried cob of purple waxy corn using response surface methodology. Journal of Food Processing and Preservation. 2018;**42**(2):e13447

[47] Yang Z, Zhai W. Optimization of microwave-assisted extraction of anthocyanins from purple corn (*Zea mays* L.) Cob and identification with HPLC – MS. Innovative Food Science and Emerging Technologies. 2010;**11**(3):470-476

[48] Chávez-González ML, Sepúlveda L, Verma DK, Luna-García HA, Rodríguez-Durán LV, Ilina A, et al. Conventional and emerging extraction processes of flavonoids. Processes. 2020;**8**(4):434

[49] Wattanathorn J, Kirisattayakul W, Suriharn B, Lertrat K. Functional drink containing the extracts of purple corn cob and Pandan leaves, the novel cognitive enhancer, increases spatial memory and hippocampal neuron density through the improvement of extracellular signal regulated protein kinase expression, Cholinergic Function, and Oxidative Status in Ovariectomized Rats. Rejuvenation Research. 2018;**21**(5):431-441

Chapter 5

Application of Liquid Chromatography in the Analysis of Flavonoid Metabolism in Plant

Ngoc Van Thi Nguyen

Abstract

Plants have evolved the capacity to create a wide range of chemicals during the process of their existence. In contrast to specialized metabolites that accumulate in a small number of plant species, flavonoids are broadly distributed across the plant kingdom. Therefore, a detailed analysis of flavonoid metabolism in genomics and metabolomics is an ideal way to investigate how plants have developed their unique metabolic pathways during the process of evolution. Among the analysis methods used for flavonoids, the coupling of liquid chromatography (LC) with ultraviolet (UV) and/or electrospray ionization (ESI) mass spectrometric detection has been demonstrated as a powerful tool for the identification and quantification of phenolics in plant extracts. This chapter mainly introduces of chemistry and metabolism of flavonoids and the application of liquid chromatography in the analysis of plant flavonoids.

Keywords: flavonoid metabolism, liquid chromatography, secondary metabolites, phenolics and phenylpropanoids, flavonoids

1. Introduction

For decades, it has been commonly accepted that plant compounds have a wide range of biological activities. They are secondary metabolites that have considerable pharmacological characteristics and play an important role in improving human health, and flavonoids are one of the substances that have been isolated. Flavonoids, which are responsible for the color and perfume of flowers, have long been known to be synthesized in specific locations, and there are presently around 6000 flavonoids that contribute to the colorful pigments of fruits, herbs, vegetables, and medicinal plants [1]. Flavonoids are hydroxylated phenolic substances known to be formed by plants in response to microbial infection and are a broad set of polyphenolic chemicals with a benzo-pyrone structure synthesized through the phenylpropanoid pathway [2–4]. Fruits, vegetables, cereals, bark, roots, stems, flowers, tea, and wine all contain it. The chemical properties of flavonoids are determined by their structural class, degree of hydroxylation, various substitutions and conjugations, and degree of polymerization [5].

Flavonoids and other phenylpropanoids are generated from phenylalanine in plants, including the subgroups of flavanones (e.g., flavanone, hesperetin, and naringenin), flavones (e.g., flavone, apigenin, and luteolin), isoflavones (e.g., daidzein, genistein, glycitein). The quantity of oxidation and pattern of substitution of the C ring change within flavonoid classes, whereas individual compounds within a class differ in the pattern of substitution of the A and B rings [6]. Flavonoids are regarded to have health-promoting effects as dietary components due to their strong antioxidant activity in both *in vivo* and *in vitro* systems [7, 8]. Flavonoids' functional hydroxyl groups influence their antioxidant activities by scavenging free radicals and/or chelating metal ions [9, 10]. In addition to antioxidant capabilities, flavonoids have been shown to have antiviral, antibacterial, anti-inflammatory, vasodilatory, anticancer, and anti-ischemic activities. The metabolism of flavonoids is performed by one or more membrane-associated multienzyme complexes rather than free-floating "soluble" enzymes [11]. The primary enzymes involved in flavonoid metabolism are chalcone synthase (a key enzyme in the phenylpropanoid pathway).

Although the separation, identification, and quantification of constituents in complex plant extracts and most likely will be a challenging task, today a multiplicity of different separation techniques, specific stationary phases, and detectors are available, helping to achieve the desired selectivity, sensitivity, and speed for nearly any separation problem. The most prominent and popular technique in this area of research is liquid chromatography [12]. Chromatographic techniques contributed significantly to the area of natural products, especially regarding identification, separation, and characterization of bioactive compounds from plant sources [13]. Flavonoid metabolism is a strong supporter in disease treatment and prevention with chemicals and is an indispensable ingredient in many nutritional, pharmaceutical, and cosmetic applications. The extensive research of flavonoid metabolism in the genome and metabolism is a great technique to investigate how plants' unique metabolic pathways originated during evolution. The coupling of liquid chromatography (LC) with ultraviolet (UV) and/or electrospray ionization (ESI) mass spectrometric detection is a potent instrument for the identification of phenolics in plant extracts. This chapter focuses on the chemistry and metabolism of flavonoids, as well as the use of liquid chromatography in the study of plant flavonoids.

2. Overview of chemistry and metabolism of flavonoid

Flavonoids are phytonutrients that belong to the polyphenol class. Polyphenols have been employed in Chinese and Ayurvedic medicine for centuries. A novel chemical was extracted from oranges in 1930. It was thought to be a member of a novel class of vitamins at the time and was labeled as vitamin P. Later, it was discovered that this chemical was a flavonoid (rutin), and over 4000 other flavonoids have been found [14].

2.1 Basic chemistry

2.1.1 General characteristics of the C_{15} unit

Flavonoids exist in the form of aglycones, glycosides, and methylated derivatives. Flavonoids have a diphenyl propane skeleton with 15 carbon atoms in their main nucleus: two six-membered rings coupled with a three-carbon unit that may or may

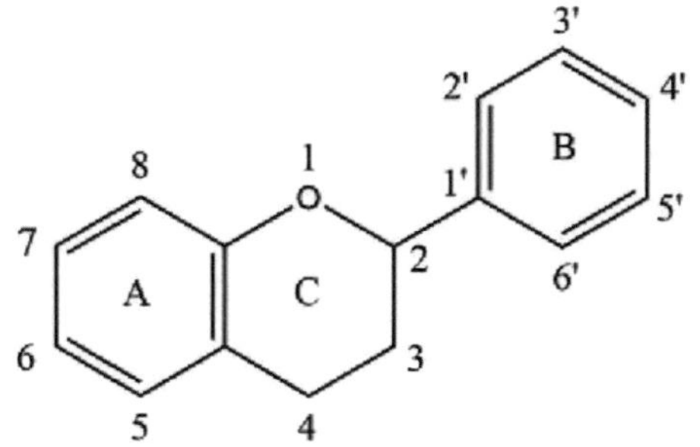

Figure 1.
General structure of flavonoid.

not be part of a third ring. Two benzene rings (A and B in **Figure 1**) are primarily connected together by a third heterocyclic oxygen-containing pyrene ring (C) [15].

Isoflavones are flavonoids in which the B ring is connected in position 3 of the C ring. Those with the B ring linked in position 4, are referred to as neoflavonoids whereas those with the B ring linked in position 2 can be further classified into many subgroups based on the structural characteristics of the C ring. Flavones, flavonols, flavanones, flavanonols, flavanols or catechins, anthocyanins, and chalcones are the subclasses [1].

2.1.2 Hydroxylation patterns of A-, B-, and C-rings

Positions 3, 5, 7, 2, 3′, 4′, and 5′ are frequently hydroxylated in flavonoids. The most prevalent A-ring hydroxylation pattern is 5,7-hydroxylation; however, a 7-hydroxy ring (also known as a 5-deoxy ring) is seen in isoflavonoid subgroups and several proanthocyanidins. On rare occasions, a 5,7,8 or 5.6.7-hydroxylation pattern is discovered. The B-ring is often 4′-, 3,4′-, or 3′,4,5′-hydroxylation. Rare flavonoids do not have B-ring oxygenation. A 2′-hydroxylation pattern is present in isoflavonoids. In isoflavonoids, the C ring is commonly hydroxylated at the carbon 3 position and occasionally at the carbon 6a position. This six-membered ring can have a carbonyl group, a hydroxyl group, a double bond between positions 2 and 3, or it can be totally unsubstituted, as in unsubstituted flavans. The isoflavonoid pterocarpans have extra rings as a consequence of 2′-hydroxylation of the original B-ring or cyclization of the added prenyl groups (**Figure 2**) [16].

2.2 Basic substitution

2.2.1 Hydroxylation

There are just a few flavonoid structures with no hydroxyl groups in the A-ring or one hydroxyl group in position 6 [17]. Such unusual structures appear to occur most frequently in the Primulaceae, Rutaceae, and Thymelaeaceae groups. However, the mechanisms of their biochemical synthesis remains unclear. The great majority of flavonoids have a basic 5,7-hydroxylation pattern of the A-ring, which is formed from malonyl-CoA during chalcone synthesis.

The C6-C3 precursor employed by chalcone synthase determines the hydroxylation pattern of the B-ring first. The physiological standard precursor is typically p-coumaroyl-CoA (4-hydroxycinnamoyl-CoA). The resultant basic C15 chalcone

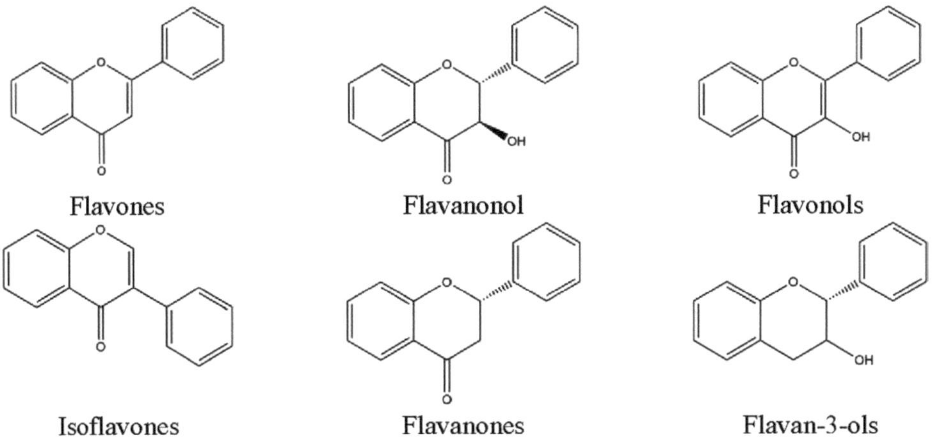

Figure 2.
Structure backbone of the main flavonoid group.

intermediate, naringenin chalcone, has a hydroxyl group in position 4′, which is seen in typical flavonoid structures, and all derived flavonoid structures have the hydroxyl group in position 4′. Thus, cinnamate 4-hydroxylase, a cytochrome P450-dependent monooxygenase that catalyzes the production of p-coumaric acid [18, 19], performs a pre-flavonoid step by introducing the hydroxyl group in position 4′.

The majority of flavonoid families have a hydroxyl group in C-ring position 3. (**Figure 1**). The well-studied flavanone 3-hydroxylase, a 2-oxoglutarate, Fe(II), and ascorbate-dependent dioxygenase [20], introduces the hydroxyl group at the flavanone level. The soluble dioxygenase catalyzes the 3-hydroxylation of the flavanone C-ring to create 3-hydroxyflavanone (flavanol). The same dioxygenase has also been linked to the catalysis of flavone synthases in several plants, with a 2-hydroxylation of the flavanone C-ring proposed as an intermediary step [16].

2.2.2 Methylation

Methylated flavonoids are a form of natural flavonoid derivative with possibly many health advantages, including enhanced bioavailability when compared to flavonoid precursors [21]. According to studies, methylating these flavonoids might boost their promise as pharmacological agents, leading to innovative uses [22]. Flavonoids have been shown to have a wide range of bioactivities, including anticancer, immunomodulation, and antioxidant activities, which can be enhanced to some extent by methylation [21]. Methylation of flavonoids through their free hydroxyl groups or C atoms significantly boosts their metabolic stability and improves membrane transport, resulting in easier absorption and significantly enhanced oral bioavailability [22]. Although cinnamic acid derivatives are methylated at the phenylpropanoid level in certain circumstances, and feruloyl-CoA serves as a poor substrate for chalcone synthase in others, methylation mainly happens at the C15 level. Poulton presents an overview of plant transmethylation and demethylation processes [23]. Grisebach documented the substrate specificities of several O-methyltransferases from parsley and soybean cell cultures, as well as shoots of *Chrysosplenium americanum* [24], while Heller and Forkmann added further test characteristics [25]. The methyl donor in all of these reactions is S-adenosyl-L-methionine. Ibrahim's group has now identified five different O-methyltransferases in the flavonol pathway from *C. americanum* [26].

2.2.3 Glycosylation

The stability of flavonoids under glycosylation reaction circumstances is an important element to consider. For some flavonoids, direct glycosylation might result in the degradation of the changed molecules. It should also be noted that glycosylation is more than just attaching the carbohydrate residue to the flavonoid component of the molecule; it also involves the removal of the protecting groups. Some flavonoids may be partially destroyed as a result of this procedure [27]. It is generally agreed that glycosylation is a late or terminal step in flavonoid glycoside biosynthesis, except for acylation and prenylation reactions. Since the glycosylation step converts the flavonoid into a more water-soluble constituent, a step necessary for the retention of some flavonoids in the vacuole, the site of the glycosylation might be expected to act at the tonoplast boundary during transfer via the cisternae of, or vesicles derived from the endoplasmic reticulum [16]. The process of direct glycosylation for some classes of flavonoids can lead to the destruction of the modified compounds. There are two major types of linkages that form either O-glycosides or C-glycosides. Parsley preparations contain both types. There is a strict specificity for the position of the hydroxyl group, generally at the 3, 5, and 7 positions of the C- and A-rings. Both 3′ and 4' B-ring glycosides are known. Recently, rarer 2′ and 5′ glycosides of highly methylated flavonol glucosides have been identified in Ibrahim's laboratory [26].

2.2.4 Acylation

Many flavonoid families include acylated sugars. They exhibit a variety of physicochemical characteristics and biological activity; however, they have limited solubility and stability. To make use of these features, various publications have indicated that enzymatic acylation of these molecules with fatty and aromatic acids by protease and lipase under varied working conditions is a potential strategy. However, it is critical to strike a balance between increasing stability and solubility while maintaining biological activity. In fact, the acylation site (regioselectivity) can significantly alter these features [28]. The acyl groups are often aromatic acids like hydroxycinnamic acids or aliphatic acids like malonic acid. They appear to be position-specific for glucoside. Malonyl glucosides, which are catalyzed by malonyl transferases, are found in isoflavonoids, flavones, flavonols, and potentially anthocyanins. O-Malonyltransferases were isolated from parsley, which included malonated flavones and flavonols. Aromatic acylation, particularly of anthocyanins, has been observed in Silene and Matthiola sp. In both cases, the acyl groups transferred were either 4-coumaroyl or caffeoyl. Acylation has been observed to promote flavonoid absorption into parsley vacuoles; alterations in the molecular symmetry of the malonylglucosides are thought to be responsible for flavonoid vacuolar entrapment inside the vacuole [29].

2.2.5 Prenylation

Prenylflavonoids are useful natural compounds found in a wide range of plants. They frequently have diverse biological features, such as phytoestrogenic, antibacterial, antitumor, and antidiabetic qualities [28]. Prenyl groups are frequently found in phytoalexins and stress-induced isoflavonoids. They are occasionally cyclized. Pterocarpans having a 2′-oxy function and a phenyl group linked to the B-ring were the most active insect feeding deterrents [30]. Elicitor-challenged bean cell cultures include a prenyltransferase in a microsomal fraction that adds a prenyl group at position 10 on the "B-ring" (also known as the D-ring) of 3,9-dihydroxypterocarpan to create phaseollidin, which is then cyclized to phaseollin. Dimethylallylpyrophosphate was the prenyl donor. The identical preparations were capable of introducing prenyl groups into medicarpin and coumestrol, but the products were not recognized. Previously, two distinct flavonoid-specific prenyltransferases that need Mn^{2+} for full activity were discovered in soybean cotyledons and cell suspension cultures [31].

2.2.6 Sulfonation

Nature mostly employs sulfation of endogenous and external substances to minimize possible harm. Sulfonated flavonoids have recently been found in substantial numbers. The majority of them are sulfate esters of common flavones and flavonols. Flavone sulfates are mostly composed of apigenin, luteolin, or its 6- and 8-hydroxy derivatives. Flavonol glycosides were sulfated through the sugar or a separate hydroxyl group. They are found in both dicots and monocots, primarily in herbaceous species or advanced morphological groupings. They are exclusively seen in ferns on rare occasions and have not been identified in bryophytes or gymnosperms [32].

2.3 Stereochemistry

Flavanones have a unique structural property known as chirality that separates them from all other groups of flavonoids (**Figure 3**). The chemical structure of all flavanones is based on a C6–C3–C6 configuration consisting of two aromatic rings connected by a three-carbon bond [33]. Almost all flavanones have one chiral carbon atom in position 2 (**Figure 3**). Except for a subgroup of flavanones known as 3-hydroxyflavanones or dihydroflavonols, which have two chiral carbon atoms in positions 2 and 3 (**Figure 4**). In the C7 position of ring A, certain flavanones include an extra d-configured mono or disaccharide sugar. These flavanone-7-O-glycosides occur as diastereoisomers or epimers with opposing configurations at just one of two or more tetrahedral stereogenic centers in the corresponding chemical entities [34].

Most natural flavonoids now only have one stereoisomer at C-2. The RS nomenclature identifies the R or S that changes at carbon 2 without any change in stereochemistry,

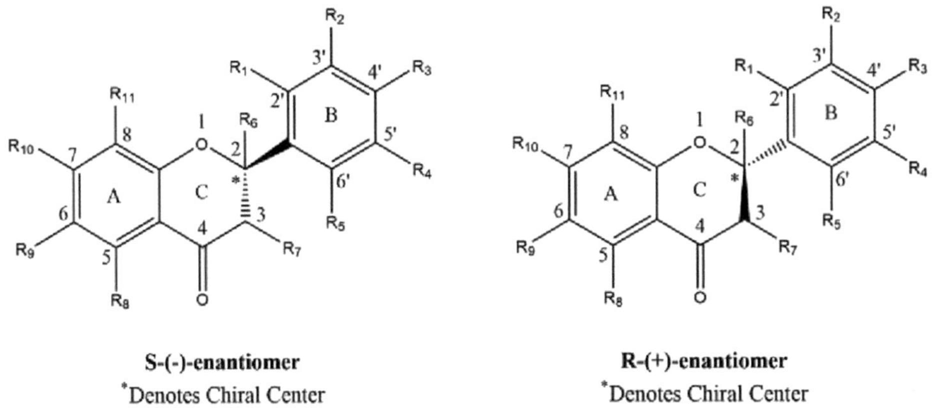

Figure 3.
Spatial disposition of the enantiomers of chiral flavanones.

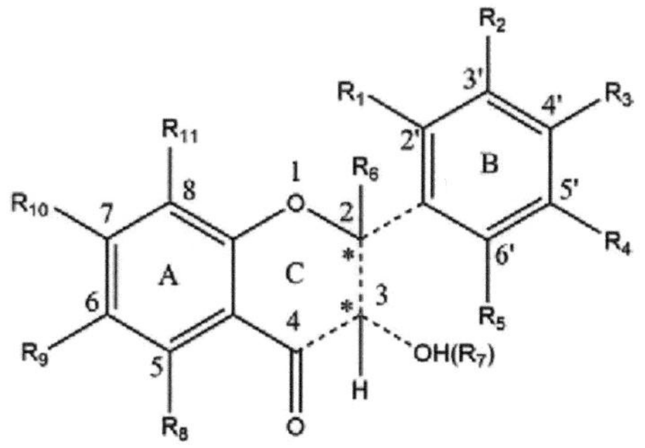

Figure 4.
Chemical structure of the chiral 3-hydroxyflavanones or dihydroflavonols.

depending on the choice of the change adjacent groups should lead to confusion for flavonoid metabolism. It is not sufficient to utilize (+) - or 2,3-cis or -trans alone to define the four potential isoforms of dihydroquercetin or catechin; consequently, consideration should be given to alternate terminology. For mirror pictures, the ent-prefix is utilized. (+)-Catechin (2,3-trans isomer) with 2R, 3S absolute stereochemistry is simply known as catechin, whilst its mirror counterpart (−)-catechin (2,3-trans) with 2R, 3R stereochemistry is simply known as ent-catechin. Similarly, the (−)-epicatechin (2,3-cis) isomer (2R, 3R) and its mirror image (2S,3S) are known as epicatechin and ent-epicatechin [35].

There are other structures designated for hydroxylation patterns and inter-liquid bonding. To minimize ambiguity in the RS system of the configuration of the interflavanoid bond at C-4, Porter and Hemingway provided sugar chemistry terminology, particularly when defining proanthocyanidin isomers. The words are also used to characterize the stereochemistry of the added hydroxyl group at the C-3 position, which results in the 2,3-cis (a-OH) and more prevalent 2,3-trans (B-OH) forms. However, such language does not accurately describe the metabolic route [16].

2.4 Overall pathways metabolism of flavonoid

Flavonoids, which include chalcones, flavones, flavonols, anthocyanins, and proanthocyanidins, are abundant in plants and have been extensively researched using biochemical and molecular biology approaches. Until recently, liverworts and mosses were thought to be the earliest flavonoid-producing plants. Genes encoding enzymes in the phenylpropanoid biosynthetic pathway, including the first two enzymes for flavonoid biosynthesis (chalcone synthase and chalcone isomerase), have not been found in the algal genera Chlamydomonas, Micromonas, Ostreococcus, and Klebsormidium, whereas genes encoding enzymes in the shikimate pathway have been found in algae, liverwort [36].

The overall route to main flavonoid groups via 5,7-hydroxy A-rings. The key intermediates in the production of flavonoids are 4-coumaroyl-coA and 3-malonyl-coA. Synthesis of 4-coumaroyl-CoA and malonyl-CoA naringenin chalcone is synthesized by chalcone synthase, an enzyme involved in the phenylpropanoid pathway. Naringenin chalcone has the ability to spontaneously cyclize to naringenin. Furthermore, naringenin chalcone synthesizes a variety of chemicals such as chalcones, aurones, and biflavonoids (it can synthesize from flavanones). Naringenin has three routes for drug synthesis. The first is the direct synthesis of isoflavonoids, followed by the addition of

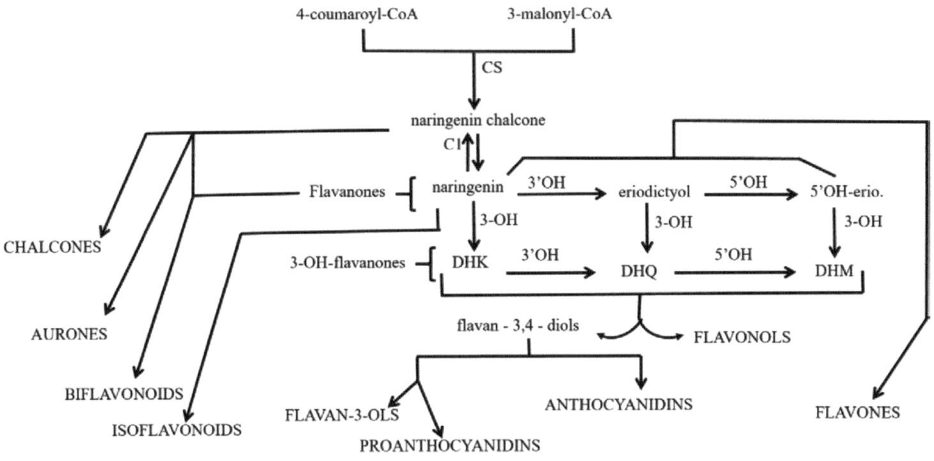

Figure 5.
Overall pathway to major flavonoid groups with 5,7-hydroxyl a rings (DHK = dihydrokaempferol; DHQ = dihydroquercetin; DHM = dihydromyricetin).

3' OH to produce eriodictyol, followed by the addition of 5'OH to produce 5'OH-erio flavones. Eriodictyol may be combined with 3-OH to get DHQ (dihydroquercetin), then with 5'OH to form DHM (dihydromyricetin) to form flavonols and flavan-3,4-diols (it can synthesize flavan-3-ols, proanthocyanidins, and anthocyanidins). Finally, adding 3-OH to naringenin results in DHK (dihydrokaempferol) (**Figure 5**) [16].

3. Liquid chromatography in the analysis of flavonoid metabolism

To show the chemical variety of flavonoids, chromatographic methods have been utilized to examine their structures. Previously, the major methods used to analyze flavonoids were paper chromatography, thin layer chromatography, column chromatography, and liquid chromatography (LC) [36]. Efficient screening of plant extracts may be accomplished using biological assays as well as chromatographic techniques such as high-performance liquid chromatography (HPLC) in conjunction with different detection modalities [37]. Because it permits systematic profiling of complex plant samples and especially focuses on their identification and consistent assessment of the found compounds, HPLC is a potent tool for the quick investigation of bioactive ingredients. Modern HPLC separation of flavonoids nearly entirely uses reversed-phase liquid chromatography (RP-LC), with significant exceptions being normal phase liquid chromatography (NP-LC) for oligomeric proanthocyanins [38] and the recent rising use of hydrophilic interaction chromatography (HILIC). Other flavonoid separation methods include mixed-mode ion-exchange-reversed phase separation of anthocyanins [39–41], size exclusions chromatography (SEC) analysis of flavonol glycosides [42], and theaflavins and proanthocyanidins [43]. However, because of the infrequent usage of the later modes, this chapter will concentrate mostly on RP-LC in line with the extent and predominance of this method in flavonoid literature.

RP-LC has proved its applicability for the separation of flavonoids depending on the nature of the aglycone (including the oxidation state, substitution patterns, and stereochemistry), the type and degree of glycosylation, and the nature and degree of acylation. The vast majority of RP-LC separations are accomplished using C18 octadecyl-silica (ODS) phases, however, C_8 [44], C_{12} [45], phenyl or phenyl-hexyl [46–50], pentafluorophenyl (PFP) [51–53] and polar embedded RP phases [54–56] as well as polymeric RP-LC phases were still widely used [57]. Aqueous/organic phases including methanol, acetonitrile, and less commonly tetrahydrofuran [58], isopropanol [59] or ethanol [57], and acidic modifiers such as acetic acid, formic acid, ammonium acetate, or trifluoroacetic acid (TFA) [59] are typical mobile phases (phosphoric, citric, or perchloric acids have also been used in combination with UV detection, although these are not suited to hyphenation with MS). Highly acidic mobile phases (>4–10% formic acid, 0.1–0.6% TFA) [60–63] are utilized for anthocyanins to assure the presence of flavylium cationic species in solution and therefore increase chromatographic efficiency. To detect and/or identify flavonoids, a variety of detectors may and have been used in conjunction with HPLC separation. These include electrochemical detection (ED) [64, 65], fluorescence (FL) [66], UV–Vis, diode array [59, 67], NMR [68, 69], and of course MS detectors [70, 71]. The most prevalent currently are diode array and MS detectors, which will be explored briefly in this and the next sections.

3.1 Liquid chromatography (LC) with ultraviolet (UV) detector

Particularly in early flavonoid research, the conjugated aromatic nature of flavonoids proved to be a significant advantage: absorption at relatively long wavelengths increases the selectivity of qualitative and quantitative spectrophotometric methods, and the distinctive spectra of various classes of flavonoids allow differentiation between them. These qualities are similarly helpful when HPLC separation is combined with UV–Vis detection. Flavonoids exhibit two UV–Vis absorption maxima: Band II (Band II), which is attributed to the A-ring, and Band III (Band III), which is attributed to the B-ring (Band 274 I). Due to the fact that it offers more specialized information and since all flavonoids absorb between 240 and 285 nm, the latter of these is more useful. Due to the absence of conjugation between the A and B rings, flavanols, flavanones, dihydroflavonols, and isoflavones only display Band II absorption (269–279 nm). Anthocyanidins may be easily identified by their Band I absorption between 460 and 550 nm in the visible range, in contrast to flavonols and flavones, which exhibit Band I absorption between 300 and 380 nm [72]. **Figure 6** provides typical illustrations of the UV–Vis absorbance spectra of the major groups of flavonoids.

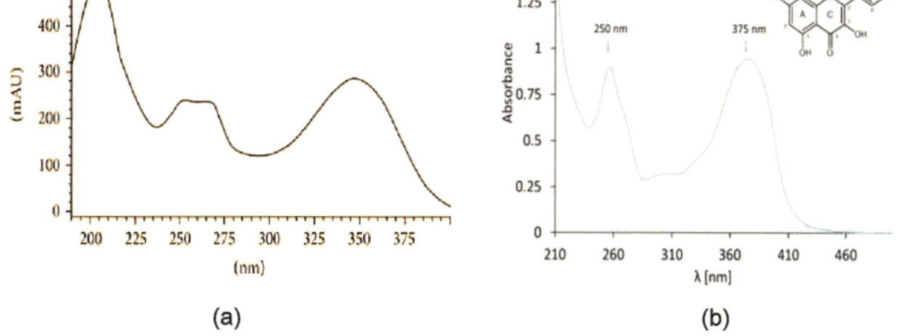

Figure 6.
UV–vis absorbance spectra of the principal classes of flavonoids: (a) Luteolin [73]; (b) quercetin [74].

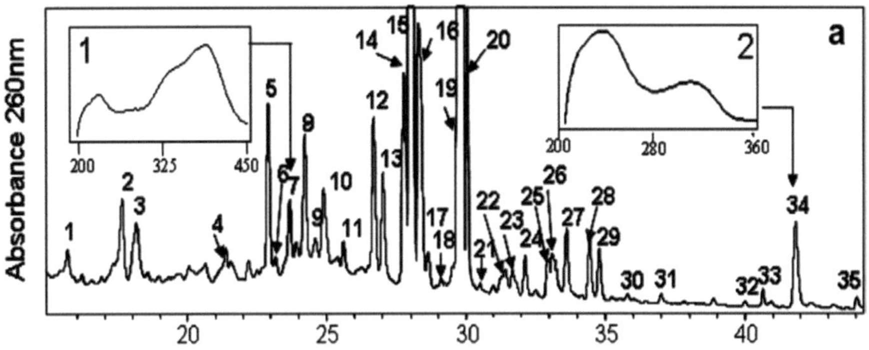

Figure 7.
HPLC–UV at 260 nm (a), 1 and 2 represent UV spectra of peak 7 (hispidol 40-O-glucoside) and peak 34 (afrormosin), respectively [75].

In research by Mohamed A. Farag et al., an integrated approach utilizing HPLC–UV was used for the large-scale and systematic identification of polyphenols in *Medicago truncatula* root and cell culture. UV spectra (200–600 nm) were recorded for different flavonoid sub-classes including 26 isoflavones (peaks 1–6, 8, 10, 12, 14, 19–29, and 31–35), 4 flavones (peaks 9, 13, 18, and 31), 2 flavanones (peaks 21 and 30), 2 aurones (peaks 7 and 11), and a chalcone (peak 32) with isoflavone representing the major sub-class. For example, isoflavones typically have a maximum absorbance near 255 nm with a second maximum between 300 and 330 nm (peak 34 in **Figure 7**), whereas aurones have the first maximum near 250 nm and the second peak around 390 nm (peak 7 in **Figure 7**) [75].

Kim-Ngan Huynh Nguyen et al.'s study quantified seven major compounds, including phenolic acids (chlorogenic acid, caffeic acid, and p-coumaric acid) and flavonoids (rutin, quercitrin, quercetin, and kaempferol) in three aerial parts of *Physalis angulata*, that is, leaves, calyces, and fruits. Chromatographic separation was carried out on a Kromasil C18 column (150 mm × 4.6 mm i.d., 5 μm) with a gradient elution of 0.1% formic acid in acetonitrile, 0.2% ammonium acetate/0.1% formic acid in water and methanol at a flow rate of 1.0 mL/min; detection was at 250 and 300 nm. The applications of liquid chromatography (LC) with ultraviolet (UV) for the analysis of flavonoid metabolism of plants that show in **Table 1** (**Figure 8**) [81].

3.2 Liquid chromatography (LC) with mass spectrometry (MS)

In contrast to 30 years ago, routine separation and preliminary identification of complex mixtures of flavonoids ranging over many orders of magnitude in concentration are now achievable because of the combination of chromatographic resolution offered by HPLC and structural data offered by MS. Furthermore, during the past 10 years, significant advancements in LC technology have been realized. UHPLC (ultra-high pressure liquid chromatography), alternative stationary phases including monoliths and superficially porous phases, high-temperature HPLC [82, 83], and multidimensional HPLC [84–87] are a few noteworthy advancements.

Cheminformatics methods combined with LC-MS/MS provide a potent tool for high-throughput surveys of flavonoid variety [88, 89]. Utilizing straightforward solvent combinations and LC columns, glycosylated, acylated, and prenylated flavonoid molecules and their aglycones may be separated. For MS/MS analysis, the isolated

Plant	Flavonoid	Detection wavelength	Stationary phases (column dimensions)	Mobile phase	Ref.
Medicago truncatula	isoflavone afrormosin, irilone	260-321 nm, 260-325 nm	C_{18}, 5 μm, 4.6 × 250 mm column	Water (0.1% acetic acid) and acetonitrile using gradient mode	[75]
Brassica rapa L. Ssp. chinensis L. (Hanelt.)	Quercetin, kaempferol, isorhamnetin	204 nm, 254 nm, 352 nm	Alltima HP C_{18} column	0.5% orthophosphoric acid (v/v) in 30% methanol (v/v) and 0.5% orthophosphoric acid (v/v) in methanol	[76]
Lupinus albus, Lupinus angustifolius	Genistein (isoflavone), 2′-hydroxygenistein glycosides	259 nm	RP C-18 silica gel	A (95% acetonitrile, 4.5% H2O, 0.5% acetic acid; v/v/v)	[77]
Brazilian Vernonieae (Asteraceae)	flavones, flavonols, flavone C-glycosides, flavonol O-glycosides	200–600 nm	Kinetex 1.7 mm XB-C18	Water and acetonitrile, both with formic acid 0.1% (v/v) using gradient mode	[78]
Sutherlandia frutescens (L.)	Four flavonoids	260 nm	C18 column	Water (0.1% acetic acid) and acetonitrile (0.1% acetic acid) using gradient mode	[79]
Orange juice	Flavanones, flavones, flavonols	280 nm, 265 nm, 265 nm	C18 standard-bore column	Aqueous formic acid (pH 2.4)/ and acetonitrile (80:20, v/v)	[80]

Table 1.
Applications of liquid chromatography (LC) with ultraviolet (UV) for the analysis of flavonoid metabolism.

molecules are ionized next. In order to analyze flavonoids, LC-tandem mass spectrometry (LC-MS/MS) has emerged as the method of choice. Algae had previously been thought to have no flavonoids. But using an LC-MS/MS technique, flavonoids were identified as intermediates and end products, demonstrating the occurrence of flavonoid production in microalgae [90]. It implies that the undiscovered flavonoids in every plant species can be discovered using cutting-edge metabolomics technology.

With an emphasis on general ionization and fragmentation processes, a brief overview of the underlying knowledge pertinent to the MS detection and MS/MS structural elucidation of flavonoids will be provided in this part. Additionally, specialized research reports provide much more in-depth information on particular classes of flavonoids, including dihydroflavonols [91], isoflavones [92], flavone-di-C-glycosides [93], flavonoid-aglycones [94], flavonoid-O-glycosides [95], and flavonoid glycosides [96]. The discussion that follows will be restricted to the API sources electrospray ionization (ESI), atmospheric pressure chemical ionization (APCI), and atmospheric pressure chemical ionization, which are presently the most pertinent LC-MS

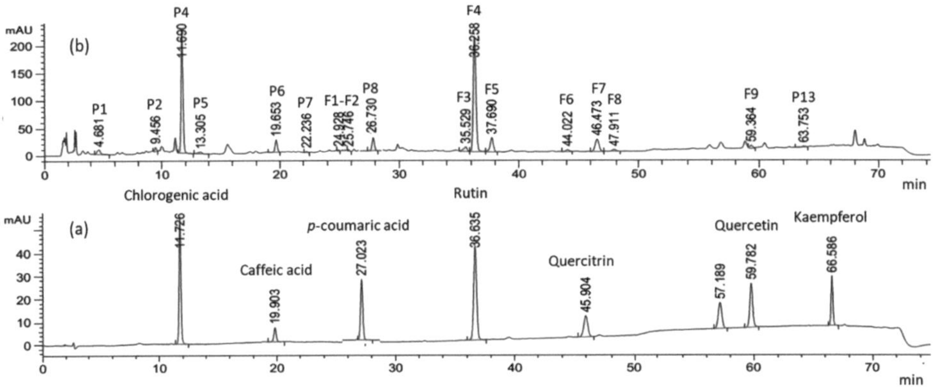

Figure 8.
HPLC chromatogram of the mixed standards solution (a), P. angulata leaves (b). P1–P13: Phenolic acids, including P4: Chlorogenic acid, P6: Caffeic acid, and P8: p-coumaric acid. F1–F9: Flavonoids, including F4: Rutin and F9: Quercetin [81].

ionization sources (APPI). The applications of liquid chromatography (LC) with mass spectrometry (MS) to the analysis of flavonoid metabolism are shown in **Table 2**.

Shoucuang Wang et al. (2017) researched comprehensive profiling of metabolites in citrus fruits. Non-targeted high-performance liquid chromatography with diode array detection and electrospray ionization mass spectrometry (HPLC-DAD-ESI-MS/MS) was used to profile the metabolites in fruit tissues. As a result, 7416 metabolic signals were detected. In addition to those reported metabolites, seven O-glycosylpolymethoxylated flavonoids were newly annotated in the study. To better characterize these flavonoids, the 3′,4′,5,6,7,8-hexamethoxyflavone standard (m70, RT 15.3 min, m/z 403.1389, error – 0.5 ppm) was analyzed first. The precursor ions of the standard compound lost one to four methyl radicals in the MS/MS spectrum to form the base peaks of [M + H - 15]+, [M + H - 30]+, [M + H - 45]+, or [M + H - 60] + (**Figure 9A**). The characteristic loss of 162 Da was observed in the MS/MS spectra corresponding to the dissociation of a hexose moiety and a series of methyl loss of the diagnostic fragments of 15 and 30 Da (**Figure 9B–D**) [102].

Paola Montoro et al. (2012) researched the metabolic profiles of different extracts (obtained by petals, stamens, and flowers) by LC-ESI-IT MS (liquid chromatography coupled with electrospray mass spectrometry equipped with an ion trap analyzer). MS/MS experiments were diagnostic for the identification of specific fragmentation patterns, that is, sugar loss for flavonoid O-glycosides or the loss of specific esterification units. Interpretation of the ESI-MS/MS experiment obtained by the analysis of *Crocus sativus* petals extracts allowed us to tentatively identify 31 flavon derivatives in the extracts under investigation, mainly glycosidated and metoxilated derivatives of kaempferol, quercetin, isorhamnetin, and tamaryxetin (**Figure 10**) [103].

3.3 High-performance liquid chromatography in chiral flavonoid

Enantiomer separation, resolution, and analysis have traditionally been achieved by the transitory or covalent synthesis of diastereoisomers. Diastereoisomers can be separated on an achiral chromatographic column by differential contact and retention because they have distinct physicochemical characteristics in an achiral environment. On a chemically bonded chiral stationary phase (CSP) with an achiral mobile phase,

Plant	Flavonoids	Stationary phase (column dimensions)	Mobile phase(s)	MS analyzer	Ref.
Arabidopsis thaliana	Chrysoeriol (3′-O-methylluteolin), kaempferol, myricetin, and quercetin	A Superspher 100 RP-18 column	A (95% acetonitrile, 4.5% H_2O, 0.5% formic acid, v/v/v) and B (95% H_2O, 4.5% acetonitrile, 0.5% formic acid v/v/v)	ESI-QqQ	[97]
Citrus fruits	Naringin, apigenin, eriodictyol, homoeriodictyol, and hesperetin	Waters BEH C_{18} column, Phenomenex Kinetex C18 column, and Agilent Poroshell 120 EC-C18 column. (2.1 mm × 50 mm, 1.7 μm)	Water and methanol, both with 0.1% formic acid	ESI-QqQ	[98]
Chenopodium hybridum	Rutin, 3-kaempferol	Eclipse XDB-C18 (150 × 4.6 mm, 5 μm	A: 0.1% Acid formic, B: Acetonitrile with 0.1% Acid formic	ESI QqQ-IT	[99]
Orange peel	Kaempferol, neohesperidin, luteolin, homoorientin, tangeretin, diosmetin formononetin quercetin, hesperidin, apigenin, naringenin, naringin, and oleuropein	Mediterranea Sea C18 (150 × 0.46 mm, 3 μm);	A: 0.1% Acid formic, B: Acetonitrile with 0.1% Acid formic	ESI-QqQ	[100]
Vitex negundo var. cannabifolia	Vitexin, hyperoside, luteoloside, liquiritin, and albiflorin	ACQUITY UPLC Cortest C18 (100 × 2.1 mm, 1.6 μm);	A: Acetonitrile, B: 5% MeOH:H2O with 0.1% acid formic	ESI-QqQ	[101]

**ESI: electrospray ionization; QqQ: triple quadrupole MS, IT: ion trap MS.*

Table 2.
Applications of liquid chromatography (LC) with mass spectrometry (MS) for the analysis of flavonoid metabolism.

racemic flavonoid resolution has typically been achieved through chromatographic enantiospecific resolution through transient production of diastereoisomers.

In 1980, one of the earliest publications on flavanone glycoside HPLC separation appeared. Both naringin and narirutin may be acetylated using an equal mixture of pyridine and acetic anhydride and then resolved at low temperatures (between 0 and 5 OC) [104]. Prunus callus (sweet cherries), oranges, and grapefruit's prunin (naringenin-7-O-glucoside) epimers were initially separated in the middle of the 1980s using benzoylated derivatives [105]. On Cyclobond I columns, the separation of

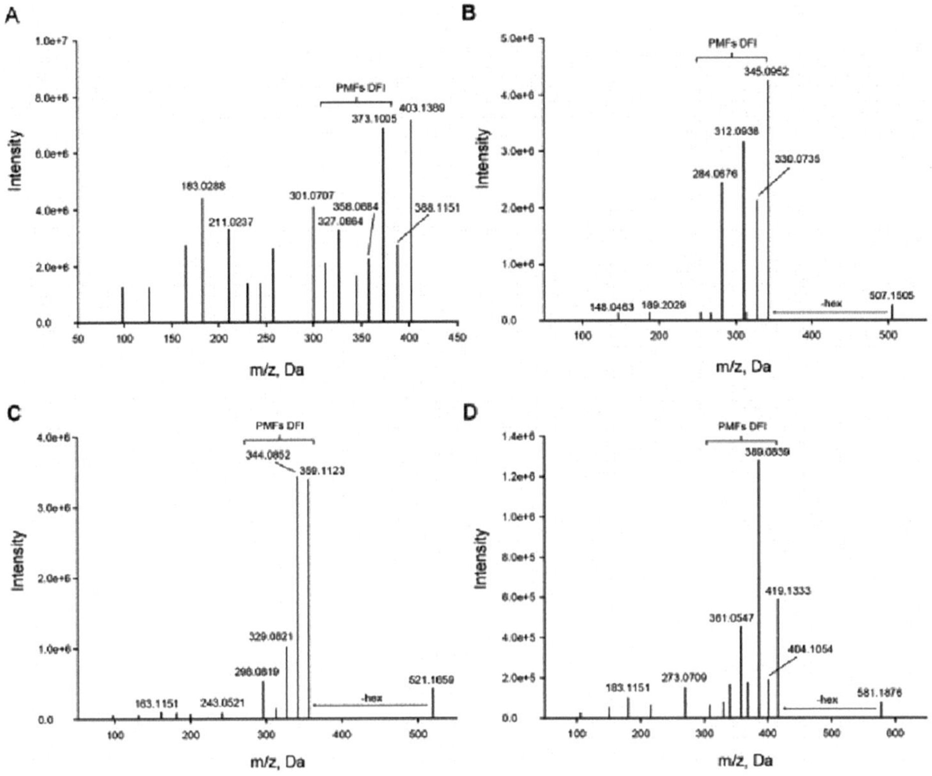

Figure 9.
Mass spectra and structures of polymethoxylated flavonoids glycosides in citrus. (A) 3, 4, 5, 6, 7,8-hexamethoxyflavone (m070). (B) Dihydroxy-trimethoxyflavone -O-hexoside (m117). (C) Hydroxytetramethoxyflavone-O-hexoside (m119). (D) Monohydroxy-hexamethoxyflavone-O-hexoside (m133). PMFs, DFI, diagnostic fragment ions of polymethoxylated flavonoids [102].

prunin benzoate and naringin benzoate has also been shown. Naringenin derivatization to naringenin tribenzoate and separation on a Chiralcel OD column are also mentioned in the literature. Naringenin's hydroxyl groups may prevent chiral identification in this stationary phase as the enantiomers could not be resolved [34].

The main advantage of chiral separation methods over achiral methods is a better understanding of the pharmacokinetics of flavanones and the development of effective dosing regimens. In the case of racemic flavanones or stereochemically pure flavanones, this requires knowledge of the *in vivo* behavior of the enantiomers and epimers. During the drug development process, understanding and comprehending the conformational stability of chiral compounds may have a significant influence on the pharmacological, pharmacokinetic, and pharmacodynamic data. Using stereospecific analytical techniques, racemization and enantiomerization/epimerization may be studied. Rapid interconversion *in vivo* would eliminate any potential distinctions in the enantiomers' medicinal or harmful effects, making the synthesis of stereochemically pure enantiomers useless. Chirality must be taken into account from the beginning of the development process for stereochemically pure compounds and racemates [34].

In a study by Gaggeri R et al. (2011), the HPLC enantioselective separation of (R/S)-naringenin (**Figure 11**), a chiral flavonoid found in several fruits juices and well-known for its beneficial health-related properties, including antioxidant, anti-inflammatory, cancer chemopreventive, immunomodulating and antimicrobial

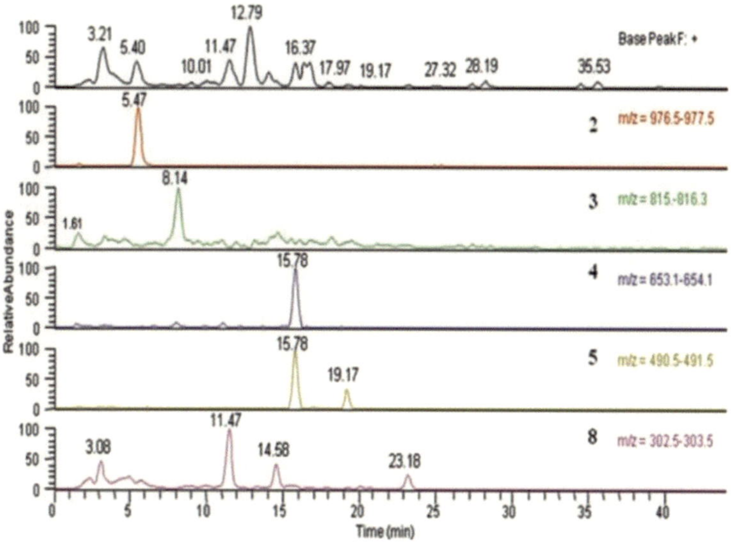

Figure 10.
LC-ESI-MS. TIC and reconstructed ion chromatograms for Crocins qualitative analysis in H2O/EtOH extract of Crocus sativus petals [103].

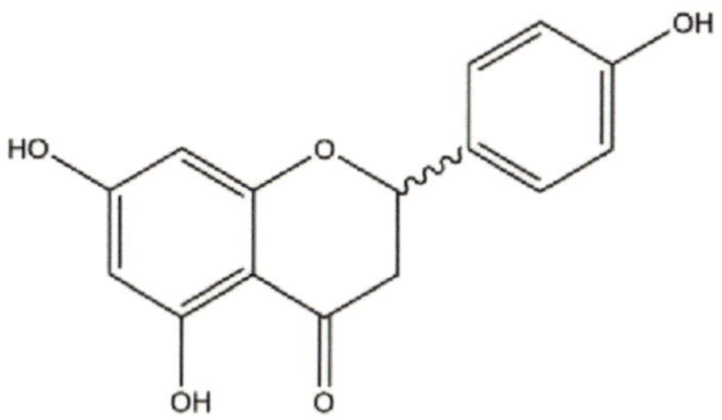

Figure 11.
(R/S)-Naringenin.

activities, has been performed on both analytical and (semi)-preparative scale using amylose-derived Chiralpak AD chiral stationary phase (CSP). A standard screening protocol for cellulose and amylose-based CSPs was firstly applied to analytical Chiralcel OD-H and Chiralpak AD-H, as well as to Lux Cellulose-1, Lux Cellulose-2, and Lux Amylose-2 in order to identify the best experimental condition for the subsequent scaling-up. Using Chiralpak AD-H and eluting with pure methanol (without acidic or basic additives), relatively short retention times, high enantioselectivity, and good resolution (Rs = 3.48) were observed. Therefore, these experimental conditions were properly scaled up to (semi)-preparative scale using both a prepacked Regispack column and a Chiralpak AD column packed in-house with bulk CSP [106].

4. Conclusion

Many beneficial health effects have been attributed to flavonoids, which are popular in the plant. The study of metabolism and bioavailability is very important in defining the pharmacological and toxicological profile of these flavonoid compounds. Due to great structural diversity among flavonoids, these profiles differ greatly from one compound to another, so the most abundant polyphenols in our diet are not necessarily the ones that reach target tissues. Therefore, careful analysis of flavonoids and their metabolites in biological systems is critical. Several hundred papers on the HPLC of flavonoids have been published in the past 20 or so years, yet HPLC methods can detect flavonoids across one, two, or perhaps three subclasses in one run. The improvements in HPLC flavonoid analysis closely resemble and, to a certain extent, build on those in domains like proteomics and metabolomics, which are supported by important breakthroughs.

Acknowledgements

The authors would like to express their hearty gratitude to Can Tho University of Medicine and Pharmacy. The authors also thank all of their colleagues for their excellent assistance.

Conflict of interest

The authors declare no conflict of interest.

Author details

Ngoc Van Thi Nguyen
Can Tho University of Medicine and Pharmacy, Can Tho City, Vietnam

*Address all correspondence to: nguyenthingocvanct@gmail.com; ntnvan@ctump.edu.vn

References

[1] Panche AN, Diwan AD, Chandra SR. Flavonoids: An overview. Journal of Nutritions Science. 2016;**5**:e47. DOI: 10.1017/jns.2016.41

[2] Mahomoodally MF, Gurib-Fakim A, Subratty AH. Screening for alternative antibiotics: An investigation into the antimicrobial activities of medicinal food plants of Mauritius. Journal of Food Science. 2010;**75**(3):M173-M177. DOI: 10.1111/j.1750-3841.2010.01555.x

[3] Kumar S, Pandey S, Pandey AK. In vitro antibacterial, antioxidant, and cytotoxic activities of Parthenium hysterophorus and characterization of extracts by LC-MS analysis. BioMed Research International. 2014;**2014**:495154. DOI: 10.1155/2014/495154

[4] Dixon RA, Dey PM, Lamb CJ. Phytoalexins: Enzymology and molecular biology. Advances in Enzymology and Related Areas of Molecular Biology. 1983;**55**:1-136. DOI: 10.1002/9780470123010.ch1

[5] Heim KE, Tagliaferro AR, Bobilya DJ. Flavonoid antioxidants: Chemistry, metabolism and structure-activity relationships. The Journal of Nutritional Biochemistry. 2002;**13**(10):572-584. DOI: 10.1016/s0955-2863(02)00208-5

[6] Middleton E Jr. Effect of plant flavonoids on immune and inflammatory cell function. Advances in Experimental Medicine and Biology. 1998;**439**:175-182. DOI: 10.1007/978-1-4615-5335-9_13

[7] Rice-Evans CA, Miller NJ, Bolwell PG, Bramley PM, Pridham JB. The relative antioxidant activities of plant-derived polyphenolic flavonoids. Free Radical Research. 1995;**22**(4):375-383. DOI: 10.3109/10715769509145649

[8] Cook NC, Samman S. Review: Flavonoids-chemistry, metabolism, cardioprotective effects and dietary sources. Journal of Nutritional Biochemistry. 1996;**7**(2):66-76

[9] Kumar S, Mishra A, Pandey AK. Antioxidant mediated protective effect of Parthenium hysterophorus against oxidative damage using in vitro models. BMC Complementary and Alternative Medicine. 2013;**13**:120. DOI: 10.1186/1472-6882-13-120

[10] Kumar S, Pandey AK. Phenolic content, reducing power and membrane protective activities of solanum xanthocarpum root extracts. Vegetos. 2013;**26**(1):301-307. DOI: 10.5958/j.2229-4473.26.1.043

[11] Winkel-Shirley B. Evidence for enzyme complexes in the phenylpropanoid and flavonoid pathways. Physiologia Plantarum. 1999;**107**(1):142-149. DOI: 10.1034/j.1399-3054.1999.100119.x

[12] Ganzera M, Sturm S. Recent advances on HPLC/MS in medicinal plant analysis-an update covering 2011-2016. Journal of Pharmaceutical and Biomedical Analysis. 2018;**147**:211-233. DOI: 10.1016/j.jpba.2017.07.038

[13] Costa DC, Costa HS, Albuquerque TG, et al. Advances in phenolic compounds analysis of aromatic plants and their potential applications. Trends in Food Science and Technology. 2015;**45**:336-354

[14] Kumar S, Pandey AK. Chemistry and biological activities of flavonoids: An overview. The Scientific World Journal. 2013;**2013**:162750. DOI: 10.1155/2013/162750

[15] Karak P. Biological activities of flavonoids: An overview. IJPSR.

2019;**3**:1567-1574. DOI: 10.13040/IJPSR.0975-8232.10(4).1567-74

[16] Stafford H. Flavonoid Metabolism. Taylor and Francis, Florida: CRC Press. 1990;**360**

[17] Harborne JB, Baxter H. A Handbook of the Natural Flavonoids. Chichester, UK: Wiley; 1999

[18] Hrazdina G, Wagner GJ. Metabolic pathways as enzyme complexes: Evidence for the synthesis of phenylpropanoids and flavonoids on membrane associated enzyme complexes. Archives of Biochemistry and Biophysics. 1985;**237**(1):88-100. DOI: 10.1016/0003-9861(85)90257-7

[19] Werck-Reichhart D. Cytochromes P450 in phenylpropanoid metabolism. Drug Metabolism and Drug Interactions. 1995;**12**(3-4):221-243. DOI: 10.1515/dmdi.1995.12.3-4.221

[20] Halbwirth H. The creation and physiological relevance of divergent hydroxylation patterns in the flavonoid pathway. International Journal of Molecular Sciences. 2010;**11**(2):595-621. DOI: 10.3390/ijms11020595

[21] Wen L, Jiang Y, Yang J, Zhao Y, Tian M, Yang B. Structure, bioactivity, and synthesis of methylated flavonoids. Annals of the New York Academy of Sciences. 2017;**1398**(1):120-129. DOI: 10.1111/nyas.13350

[22] Koirala N, Thuan NH, Ghimire GP, Thang DV, Sohng JK. Methylation of flavonoids: Chemical structures, bioactivities, progress and perspectives for biotechnological production. Enzyme and Microbial Technology. 2016;**86**:103-116. DOI: 10.1016/j.enzmictec.2016.02.003

[23] Poulton JE. Transmethylation and demethylation reac tions in the metabolism of secondary plant products. In: Conn EE, editor. The Biochemistry of Plants, Vol. 7: Secondary Plant Products. New York: Academic Press; 1981. pp. 667-723

[24] De Luca V, Ibrahim RK. Enzymatic synthesis of polymethylated flavonols in Chrysosplenium americanum. I. Partial purification and some properties of S-adenosyl-L-methionine: Flavonol 3-, 6-, 7-, and 4'-O-methyltransferases. Archives of Biochemistry and Biophysics. 1985;**238**(2):596-605. DOI: 10.1016/0003-9861(85)90205-x

[25] Hahlbrock K, Grisebach H. Biosynthesis of Flavonoids. Boston, MA: In The flavonoids Springer; 1975. pp. 866-915

[26] Ibrahim RK, De Luca V, Khouri H, Latchinian L, Brisson L, Charest PM. Enzymology and compartmentation of polymethylated flavonol glucosides in Chrysosplenium americanum. Phytochemistry. 1987;**26**(5):1237-1245

[27] Khodzhaieva RS, Gladkov ES, Kyrychenko A, Roshal AD. Progress and achievements in glycosylation of flavonoids. Frontiers in Chemistry. 2021;**9**:637994. Published 2021 Mar 31. DOI: 10.3389/fchem.2021.637994

[28] Yang X, Yang J, Jiang Y, et al. Regiospecific synthesis of prenylated flavonoids by a prenyltransferase cloned from fusarium oxysporum. Scientific Reports. 2016;**6**:24819. Published 2016 Apr 21. DOI: 10.1038/srep24819

[29] Matern U, Reichenbach C, Heller W. Efficient uptake of flavonoids into parsley (Petroselinum hortense) vacuoles requires acylated glycosides. Planta. 1986;**167**(2):183-189. DOI: 10.1007/BF00391413

[30] Yoneyama K, Akashi T, Aoki T. Molecular characterization of soybean

Pterocarpan 2-Dimethylallyltransferase in Glyceollin biosynthesis: Local gene and Whole-genome duplications of Prenyltransferase genes led to the structural diversity of soybean Prenylated Isoflavonoids. Plant & Cell Physiology. 2016;**57**(12):2497-2509. DOI: 10.1093/pcp/pcw178

[31] Zähringer U, Schaller E, Grisebach H. Induction of phytoalexin synthesis in soybean. Structure and reactions of naturally occurring and enzymatically prepared prenylated pterocarpans from elicitor-treated cotyledons and cell cultures of soybean. Zeitschrift für Naturforschung C. 1981;**36**(3-4):234-241

[32] Correia-da-Silva M, Sousa E, Pinto MM. Emerging sulfated flavonoids and other polyphenols as drugs: Nature as an inspiration. Medicinal Research Reviews. 2014;**34**(2):223-279. DOI: 10.1002/med.21282

[33] Shahidi F, Wanasundara PK. Phenolic antioxidants. Critical Reviews in Food Science and Nutrition. 1992;**32**(1):67-103. DOI: 10.1080/10408399209527581

[34] Yáñez JA, Andrews PK, Davies NM. Methods of analysis and separation of chiral flavonoids. Journal of Chromatography. B, Analytical Technologies in the Biomedical and Life Sciences. 2007;**848**(2):159-181. DOI: 10.1016/j.jchromb.2006.10.052

[35] Hemingway RW, Foo LY, Porter LJ. Linkage isomerism in trimeric and polymeric 2,3-cis-procyanidins. Journal of the Chemical Society, Perkin Transactions. 1982;**1**:1209

[36] Yonekura-Sakakibara K, Higashi Y, Nakabayashi R. The origin and evolution of plant flavonoid metabolism. Frontiers in Plant Science. 2019;**10**:943. Published 2019 Aug 2. DOI: 10.3389/fpls.2019.00943

[37] Wolfender JL, Rodriguez S, Hostettmann K. Liquid chromatography coupled to mass spectrometry and nuclear magnetic resonance spectroscopy for the screening of plant constituents. Journal of Chromatography. A. 1998;**794**(1-2):299-316. DOI:10.1016/s0021-9673(97)00939-4

[38] LEA, Andrew GH. High performance liquid chromatography of cider procyanidins. Journal of the Science of Food and Agriculture. 1979;**30**(8):833-838

[39] McCallum JL, Yang R, Young JC, Strommer JN, Tsao R. Improved high performance liquid chromatographic separation of anthocyanin compounds from grapes using a novel mixed-mode ion-exchange reversed-phase column. Journal of Chromatography. A. 2007;**1148**(1):38-45. DOI: 10.1016/j.chroma.2007.02.088

[40] Vergara C, Mardones C, Hermosín-Gutiérrez I, von Baer D. Comparison of high-performance liquid chromatography separation of red wine anthocyanins on a mixed-mode ion-exchange reversed-phase and on a reversed-phase column. Journal of Chromatography. A. 2010;**1217**:5710-5717

[41] Jin H, Liu Y, Guo Z, Yang F, Wang J, Li X, et al. High-performance liquid chromatography separation of cis-trans anthocyanin isomers from wild Lycium ruthenicum Murr. Employing a mixed-mode reversed-phase/strong anion-exchange stationary phase. Journal of Agricultural and Food Chemistry. 2015;**63**(2):500-508. DOI: 10.1021/jf504525w

[42] de Souza LM, Cipriani TR, Sant'ana CF, Iacomini M, Gorin PA, Sassaki GL. Heart-cutting two-dimensional (size exclusion x reversed phase) liquid chromatography-mass spectrometry

analysis of flavonol glycosides from leaves of Maytenus ilicifolia. Journal of Chromatography. A. 2009;**1216**(1):99-105. DOI: 10.1016/j.chroma.2008.11.032

[43] Scoparo CT, de Souza LM, Dartora N, Sassaki GL, Gorin PA, Iacomini M. Analysis of Camellia sinensis green and black teas via ultra high performance liquid chromatography assisted by liquid-liquid partition and two-dimensional liquid chromatography (size exclusion × reversed phase). Journal of Chromatography. A. 2012;**1222**:29-37. DOI: 10.1016/j.chroma.2011.11.038

[44] Harborne JB, Boardley M. Use of high-performance liquid chromatography in the separation of flavonol glycosides and flavonol sulphates. Journal of Chromatography A. 1984;**299**:377-385. DOI: 10.1016/s0021-9673(01)97853-7

[45] Omar MH, Mullen W, Crozier A. Identification of proanthocyanidin dimers and trimers, flavone C-glycosides, and antioxidants in Ficus deltoidea, a Malaysian herbal tea. Journal of Agricultural and Food Chemistry. 2011;**59**(4):1363-1369

[46] Dugo P, Cacciola F, Herrero M, Donato P, Mondello L. Use of partially porous column as second dimension in comprehensive two-dimensional system for analysis of polyphenolic antioxidants. Journal of Separation Science. 2008;**31**(19):3297-3308. DOI: 10.1002/jssc.200800281

[47] Tanaka Y, Yanagida A, Komeya S, Kawana M, Honma D, Tagashira M, et al. Comprehensive separation and structural analyses of polyphenols and related compounds from bracts of hops (Humulus lupulus L.). Journal of Agricultural and Food Chemistry. 2014;**62**(10):2198-2206. DOI: 10.1021/jf405544n

[48] Cesla P, Fischer J, Jandera P. Separation of phenolic acids and flavone natural antioxidants by two-dimensional method combining liquid chromatography and micellar electrokinetic capillary chromatography. Electrophoresis. 2010;**31**(13):2200-2210. DOI: 10.1002/elps.200900689

[49] Guillarme D, Casetta C, Bicchi C, Veuthey JL. High throughput qualitative analysis of polyphenols in tea samples by ultra-high pressure liquid chromatography coupled to UV and mass spectrometry detectors. Journal of Chromatography. A. 2010;**1217**(44):6882-6890. DOI: 10.1016/j.chroma.2010.08.060

[50] Wu LZ, Zhang XP, Xu XD, Zheng QX, Yang JS, Ding WL. Characterization of aromatic glycosides in the extracts of Trollius species by ultra high-performance liquid chromatography coupled with electrospray ionization quadrupole time-of-flight tandem mass spectrometry. Journal of Pharmaceutical and Biomedical Analysis. 2013;**75**:55-63

[51] Jandera P, Hájek T, Vyňuchalová K. Retention and bandwidths prediction in fast gradient liquid chromatography. Part 2-Core-shell columns. Journal of Chromatography. A. 2014;**1337**:57-66

[52] Mirali M, Ambrose SJ, Wood SA, Vandenberg A, Purves RW. Development of a fast extraction method and optimization of liquid chromatography-mass spectrometry for the analysis of phenolic compounds in lentil seed coats. Journal of Chromatography. B, Analytical Technologies in the Biomedical and Life Sciences. 2014;**969**:149-161. DOI: 10.1016/j.jchromb.2014.08.007

[53] Gavina JM, Priem J, Wood CM, Xiao CW, Feng YL. Determination

of isoflavones in rat serum using liquid chromatography-tandem mass spectrometry with a highly efficient core-shell column. Analytical and Bioanalytical Chemistry. 2013;**405**(8):2643-2651. DOI: 10.1007/s00216-012-6688-x

[54] Feuereisen MM, Hoppe J, ZimmermannBF,WeberF,Schulze-KaysersN, Schieber A. Characterization of phenolic compounds in Brazilian pepper (Schinus terebinthifolius Raddi) exocarp. Journal of Agricultural and Food Chemistry. 2014;**62**(26):6219-6226. DOI: 10.1021/jf500977d

[55] Li C, Zhao Y, Guo Z, Zhang X, Xue X, Liang X. Effective 2D-RPLC/RPLC enrichment and separation of micro-components from Hedyotis diffusa Willd. And characterization by using ultra-performance liquid chromatography/quadrupole time-of-flight mass spectrometry. Journal of Pharmaceutical and Biomedical Analysis. 2014;**99**:35-44. DOI: 10.1016/j.jpba.2014.06.020

[56] Šatínský D, Jägerová K, Havlíková L, et al. A new and fast HPLC method for determination of Rutin, Troxerutin, Diosmin and hesperidin in food supplements using fused-Core column technology. Food Analytical Methods. 2013;**6**:1353-1360. DOI: 10.1007/s12161-012-9551-y

[57] Reichelt KV, Peter R, Paetz S, Roloff M, Ley JP, Krammer GE, et al. Characterization of flavor modulating effects in complex mixtures via high temperature liquid chromatography. Journal of Agricultural and Food Chemistry. 2010;**58**(1):458-464. DOI: 10.1021/jf9027552

[58] Olszewska MA. New validated high-performance liquid chromatographic method for simultaneous analysis of ten flavonoid aglycones in plant extracts using a C18 fused-core column and acetonitrile-tetrahydrofuran gradient. Journal of Separation Science. 2012;**35**(17):2174-2183. DOI: 10.1002/jssc.201200287

[59] Santos-Buelga C, García-Viguera C, Tomas-Barberan FA. On-line identification of flavonoids by HPLC coupled to diode Array detection, in: C. Santos-Buelga, G. Williamson (Eds.). Methods Polyphen. Anal., The Royal Society of Chemistry, Cambridge, 2003: pp. 92-127.

[60] Lopes-da-Silva F, de Pascual-Teresa S, Rivas-Gonzalo J, Santos-Buelga C. Identification of anthocyanin pigments in strawberry (cv Camarosa) by LC using DAD and ESI-MS detection. European Food Research and Technology. 2002;**214**(3):248-253

[61] de Villiers A, Cabooter D, Lynen F, Desmet G, Sandra P. High performance liquid chromatography analysis of wine anthocyanins revisited: Effect of particle size and temperature. Journal of Chromatography. A. 2009;**1216**(15):3270-3279. DOI: 10.1016/j.chroma.2009.02.038

[62] Xu Y, Simon JE, Ferruzzi MG, Ho L, Pasinetti GM, Wu Q. Quantification of anthocyanidins in the grapes and grape juice products with acid assisted hydrolysis using LC/MS. Journal of Functional Foods. 2012;**4**:710-717

[63] Díaz-García MC, Obón JM, Castellar MR, Collado J, Alacid M. Quantification by UHPLC of total individual polyphenols in fruit juices. Food Chemistry. 2013;**138**(2-3):938-949. DOI: 10.1016/j.foodchem.2012.11.061

[64] Manach C. The use of HPLC with Coulometric detection in the analysis of flavonoids in complex matrices. In: Santos-Buelga C, Williamson G, editors.

Methods Polyphen. Anal. Cambridge: The Royal Society of Chemistry; 2003. pp. 63-91

[65] Milbury PE. Analysis of complex mixtures of flavonoids and polyphenols by high-performance liquid chromatography electrochemical detection methods. Methods in Enzymology. 2001;**335**:15-26. DOI: 10.1016/s0076-6879(01)35228-x

[66] Wolfender JL. HPLC in natural product analysis: The detection issue. Planta Medica. 2009;**75**(7):719-734. DOI: 10.1055/s-0028-1088393

[67] Lin LZ, Harnly JM. Quantitation of flavanols, proanthocyanidins, isoflavones, flavanones, dihydrochalcones, stilbenes, benzoic acid derivatives using ultraviolet absorbance after identification by liquid chromatography-mass spectrometry. Journal of Agricultural and Food Chemistry. 2012;**60**(23):5832-5840. DOI: 10.1021/jf3006905

[68] Rupasinghe HPV. Application of NMR spectroscopy in plant polyphenols associated with human health. Applications of NMR spectroscopy. Netherlands: Bentham Science Publishers. 2015:3-92.

[69] Wolfender JL, Marti G, Thomas A, Bertrand S. Current approaches and challenges for the metabolite profiling of complex natural extracts. Journal of Chromatography. A. 2015;**1382**:136-164. DOI: 10.1016/j.chroma.2014.10.091

[70] de Rijke E, Out P, Niessen WM, Ariese F, Gooijer C, Brinkman UA. Analytical separation and detection methods for flavonoids. Journal of Chromatography. A. 2006;**1112**(1-2):31-63. DOI: 10.1016/j.chroma.2006.01.019

[71] de Pascual-Teresa S. Rivas-Gonzalo JC. Application of LC-MS for the identification of polyphenols. In: Santos-Buelga C, Williamson G, editors. Methods Polyphen. Anal. Cambridge: The Royal Society of Chemistry; 2003. pp. 48-62

[72] de Villiers A, Venter P, Pasch H. Recent advances and trends in the liquid-chromatography-mass spectrometry analysis of flavonoids. Journal of Chromatography. A. 2016;**1430**:16-78. DOI: 10.1016/j.chroma.2015.11.077

[73] da Silva JB, Temponi Vdos S, Gasparetto CM, Fabri RL, Aragão DM, Pinto Nde C, et al. Vernonia condensata baker (Asteraceae): A promising source of antioxidants. Oxidative Medicine and Cellular Longevity. 2013;**2013**:698018. DOI: 10.1155/2013/698018

[74] Golonka I, Wilk S, Musiał W. The influence of UV radiation on the degradation of pharmaceutical formulations containing quercetin. Molecules. 2020;**25**(22):5454. DOI: 10.3390/molecules25225454

[75] Farag MA, Huhman DV, Lei Z, Sumner LW. Metabolic profiling and systematic identification of flavonoids and isoflavonoids in roots and cell suspension cultures of Medicago truncatula using HPLC-UV-ESI-MS and GC-MS. Phytochemistry. 2007;**68**(3):342-354

[76] Simone J. Rochfort, Michael Imsic, Rod Jones, V. Craige Trenerry, And Bruce Tomkins (2006) .Characterization of Flavonol conjugates in immature leaves of Pak Choi [Brassica rapa L. Ssp. chinensis L. (Hanelt.)] by HPLC-DAD and LC-MS/MS. Journal of Agricultural and Food Chemistry, 54, pp. 4855-4860.

[77] Bednarek P, Kerhoas L, Einhorn J, Frański R, Wojtaszek P, Rybus-Zajac M, et al. Profiling of flavonoid conjugates in Lupinus albus and Lupinus

angustifolius responding to biotic and abiotic stimuli. Journal of Chemical Ecology. 2003;**29**(5):1127-1142. DOI: 10.1023/a:1023877422403

[78] Gallon ME, Jaiyesimi OA, Gobbo-Neto L. LC-UV-HRMS dereplication of secondary metabolites from Brazilian Vernonieae (Asteraceae) species supported through in-house database. Biochemical Systematics and Ecology. 2018;**78**:5-16

[79] Avula B, Wang YH, Smillie TJ, Fu X, Li XC, Mabusela W, et al. Quantitative determination of flavonoids and cycloartanol glycosides from aerial parts of Sutherlandia frutescens (L.) R. BR. By using LC-UV/ELSD methods and confirmation by using LC-MS method. Journal of Pharmaceutical and Biomedical Analysis. 2010;**52**(2):173-180. DOI: 10.1016/j.jpba.2010.01.010

[80] Careri M, Elviri L, Mangia A, Musci M. Spectrophotometric and coulometric detection in the high-performance liquid chromatography of flavonoids and optimization of sample treatment for the determination of quercetin in orange juice. Journal of Chromatography. A. 2000;**881**(1-2):449-460. DOI: 10.1016/s0021-9673(00)00256-9

[81] Nguyen K-NH, Nguyen N-VT, Kim KH. Determination of phenolic acids and flavonoids in leaves, calyces, and fruits of Physalis angulata L. in Viet Nam. Pharmacia. 2021;**68**(2):501-509. DOI: 10.3897/pharmacia.68.e66044

[82] Guillarme D, Ruta J, Rudaz S, Veuthey JL. New trends in fast and high-resolution liquid chromatography: A critical comparison of existing approaches. Analytical and Bioanalytical Chemistry. 2010;**397**(3):1069-1082. DOI: 10.1007/s00216-009-3305-8

[83] Fekete S, Veuthey JL, Guillarme D. Comparison of the most recent chromatographic approaches applied for fast and high resolution separations: Theory and practice. Journal of Chromatography. A. 2015;**1408**:1-14. DOI: 10.1016/j.chroma.2015.07.014

[84] Stoll DR, Li X, Wang X, Carr PW, Porter SE, Rutan SC. Fast, Comprehensive two-dimensional liquid chromatography. Journal of Chromatography. A 2007;1168(1-2):3-43; discussion 2. DOI: 10.1016/j.chroma.2007.08.054

[85] François I, Sandra K, Sandra P. Comprehensive liquid chromatography: Fundamental aspects and practical considerations--a review. Analytica Chimica Acta. 2009;**641**(1-2):14-31. DOI: 10.1016/j.aca.2009.03.041

[86] Stoll DR. Recent progress in online, comprehensive two-dimensional high-performance liquid chromatography for non-proteomic applications. Analytical and Bioanalytical Chemistry. 2010;**397**(3):979-986. DOI: 10.1007/s00216-010-3659-y

[87] Malerod H, Lundanes E, Greibrokk T. Recent advances in on-line multidimensional liquid chromatography. Analytical Methods. 2010;**2**:110-122

[88] Akimoto N, Ara T, Nakajima D, Suda K, Ikeda C, Takahashi S, et al. Flavonoid search: A system for comprehensive flavonoid annotation by mass spectrometry. Scientific Reports. 2017;**7**(1):1243. DOI: 10.1038/s41598-017-01390-3

[89] Tsugawa H, Kind T, Nakabayashi R, Yukihira D, Tanaka W, Cajka T, et al. Hydrogen rearrangement rules: Computational MS/MS fragmentation and structure elucidation using MS-FINDER software. Analytical

Chemistry. 2016;**88**(16):7946-7958. DOI: 10.1021/acs.analchem.6b00770

[90] Goiris K, Muylaert K, Voorspoels S, Noten B, De Paepe D, Baart E, et al. Detection of flavonoids in microalgae from different evolutionary lineages. Journal of Phycology. 2014;**50**(3):483-492. DOI: 10.1111/jpy.12180

[91] Abad-García B, Garmón-Lobato S, Berrueta LA, Gallo B, Vicente F. A fragmentation study of dihydroquercetin using triple quadrupole mass spectrometry and its application for identification of dihydroflavonols in citrus juices. Rapid Communications in Mass Spectrometry. 2009;**23**(17):2785-2792. DOI: 10.1002/rcm.4182

[92] Vacek J, Klejdus B, Lojková L, Kubán V. Current trends in isolation, separation, determination and identification of isoflavones: A review. Journal of Separation Science. 2008;**31**(11):2054-2067. DOI: 10.1002/jssc.200700569

[93] Abad-García B, Garmón-Lobato S, Berrueta LA, Gallo B, Vicente F. New features on the fragmentation and differentiation of C-glycosidic flavone isomers by positive electrospray ionization and triple quadrupole mass spectrometry. Rapid Communications in Mass Spectrometry. 2008;**22**(12):1834-1842. DOI: 10.1002/rcm.3560

[94] Hughes RJ, Croley TR, Metcalfe CD, March RE. A tandem mass spectrometric study of selected characteristic flavonoids. International Journal of Mass Spectrometry. 2001;**211**:371-385

[95] Cuyckens F, Rozenberg R, de Hoffmann E, Claeys M. Structure characterization of flavonoid O-diglycosides by positive and negative nano-electrospray ionization ion trap mass spectrometry. Journal of Mass Spectrometry. 2001;**36**(11):1203-1210. DOI: 10.1002/jms.224

[96] March RE, Lewars EG, Stadey CJ, Miao X-S, Zhao X, Metcalfe CD. A comparison of flavonoid glycosides by electrospray tandem mass spectrometry. International Journal of Mass Spectrometry. 2006;**248**:61-85

[97] Kachlicki P, Einhorn J, Muth D, Kerhoas L, Stobiecki M. Evaluation of glycosylation and malonylation patterns in flavonoid glycosides during LC/MS/MS metabolite profiling. Journal of Mass Spectrometry. 2008;**43**(5):572-586. DOI: 10.1002/jms.1344

[98] Zeng X, Zheng Y, He Y, Peng W, Su W. A rapid LC-MS/MS method for simultaneous determination of ten flavonoid metabolites of Naringin in rat urine and its application to an excretion study. Food. 2022;**11**:316. DOI: 10.3390/foods11030316

[99] Podolak I, Olech M, Galanty A, Załuski D, Grabowska K, Sobolewska D, et al. Flavonoid and phenolic acid profile by LC-MS/MS and biological activity of crude extracts from Chenopodium hybridum aerial parts. Natural Product Research. 2016;**30**(15):1766-1770. DOI: 10.1080/14786419.2015.1136908

[100] Molina-Calle M, Priego-Capote F, Luque de Castro MD. Development and application of a quantitative method for determination of flavonoids in orange peel: Influence of sample pretreatment on composition. Talanta. 2015;**144**:349-355. DOI: 10.1016/j.talanta.2015.05.054

[101] Huang M, Zhang Y, Xu S, Xu W, Chu K, Xu W, et al. Identification and quantification of phenolic compounds in Vitex negundo L. var. cannabifolia (Siebold et Zucc.) hand.-Mazz. Using

liquid chromatography combined with quadrupole time-of-flight and triple quadrupole mass spectrometers. Journal of Pharmaceutical and Biomedical Analysis. 2015;**108**:11-20. DOI: 10.1016/j.jpba.2015.01.049

[102] Wang S, Yang C, Tu H, et al. Characterization and metabolic diversity of flavonoids in citrus species. Scientific Reports. 2017;7(1):10549. Published 2017. DOI: 10.1038/s41598-017-10970-2

[103] Montoro P, Maldini M, Luciani L, Tuberoso CI, Congiu F, Pizza C. Radical scavenging activity and LC-MS metabolic profiling of petals, stamens, and flowers of Crocus sativus L. Journal of Food Science. 2012;77(8):C893-C900. DOI: 10.1111/j.1750-3841.2012.02803.x

[104] Rudolf G, Herrmann K. Analysis of flavonoids by high-performance liquid chromatography. Journal of Chromatography A. 1980;**189**(2):217-224. DOI: 10.1016/s0021-9673(00)81521-6

[105] Feucht W, Treutter D, Schmid P. Inhibition of growth and xylogenesis and promotion of vacuolation in prunus callus by the flavanone prunin. Plant Cell Reports. 1988;7(3):189-192. DOI: 10.1007/BF00269320

[106] Gaggeri R, Rossi D, Collina S, Mannucci B, Baierl M, Juza M. Quick development of an analytical enantioselective high performance liquid chromatography separation and preparative scale-up for the flavonoid Naringenin. Journal of Chromatography. A. 2011;**1218**(32):5414-5422. DOI: 10.1016/j.chroma.2011.02.038

Chapter 6

Recent Advances in Flavonoid Metabolism: An Updated Review

Indireddy Theja and Banoth Ramya Kuber

Abstract

Flavonoids are polyhydroxylated natural chemicals that have been shown to improve human health. These are a type of bioactive molecules that can be found in abundance in plants. These polyphenolic chemicals are naturally generated from plant metabolites. Before entering the intestine, flavonoid glycosides are deglycosylated, while aglycones can readily pass-through cell membranes. They are absorbed and transferred to the liver, where they undergo substantial metabolism, resulting in glucuronides, sulfates, and methylation compounds. These conjugates are responsible for the health-promoting possessions of flavonoids. The flavonol subclass was the first to be researched, with quercetin as the most common dietary flavonol, and information on other flavonoid subclasses is still developing. Cellular signaling pathways mediate the antidiabetic benefits of dietary flavonoids in the pancreas, liver, and skeletal muscle. Flavonoids modulate distinct signaling pathways in pancreatic cells, hepatocytes, adipocytes, and skeletal myofibers via acting on various molecular targets. Flavonoids may help people with diabetes firstly by improving hyperglycemia through glucose metabolism regulation in hepatocytes and secondly by reducing insulin resistance, inflammation, and oxidative stress in muscle and fat and by increasing glucose uptake in skeletal muscle and white adipose tissue. A greater understanding of the flavonoid pathway's regulatory mechanisms would likely favor the progress of novel bioprocessing techniques for the production of value-added plants with optimal flavonoid content.

Keywords: deglycosylated, hyperglycemia, molecular targets, plant metabolites, polyhydroxylated natural chemicals

1. Introduction

Flavonoids are secondary metabolites found in high concentrations in vascular florae and minor amounts in lichens. They get to build up in all structures and matters at various periods of expansion and in response to conservational factors. These compounds are of prodigious significance to social nourishment and wellbeing, and their many characteristics in plant progress and variation in the atmosphere. They do help in the organoleptic eminence of plant-derived goods, as well as being helpful to social wellbeing and cell aging anticipation. Increased consumption of vegetables and fruits has been known to protect against cancer and cardiovascular disease. These are a major class of natural antioxidants found in a plant-based diet and may have a role in this protective effect. Fruits (cherry, grapes, and apple), vegetables (onion, broccoli, and spinach), beverages (coffee and tea), soy products, and basils are all good sources of flavonoids [1]. They can be found inside the cells of all plant tissues as well as on the exteriors of many plant tissues. The phenylpropanoid unit, C6C3, is a common building component in the carbon skeleton of these phenols. This route creates a wide range of plant phenols during biosynthesis. The molecular structure of this class of composites is created on a diphenyl propane (C6-C3-C6) skeleton with 2 aromatic rings joined by a 3-carbon "bridge" to form a 6-member heterocyclic ring. Flavonoids are classified into 3 groups based on the aromatic ring's link to the heterocyclic ring: flavonoids (2-phenylbenzopyrans), isoflavonoids (3-phenylbenzopyrans), and neoflavonoids (4-phenylbenzopyrans) [2]. They can be separated into numerous sets based on the degree of oxidation and saturation in the heterocyclic C-ring, as shown in **Figure 1**. Hydroxylation occurs in positions 3, 5, 7, 3′, 4′, and/or 5′ in flavonoids. Methylated, acylated, prenylated, or sulfated hydroxyl groups are frequently found.

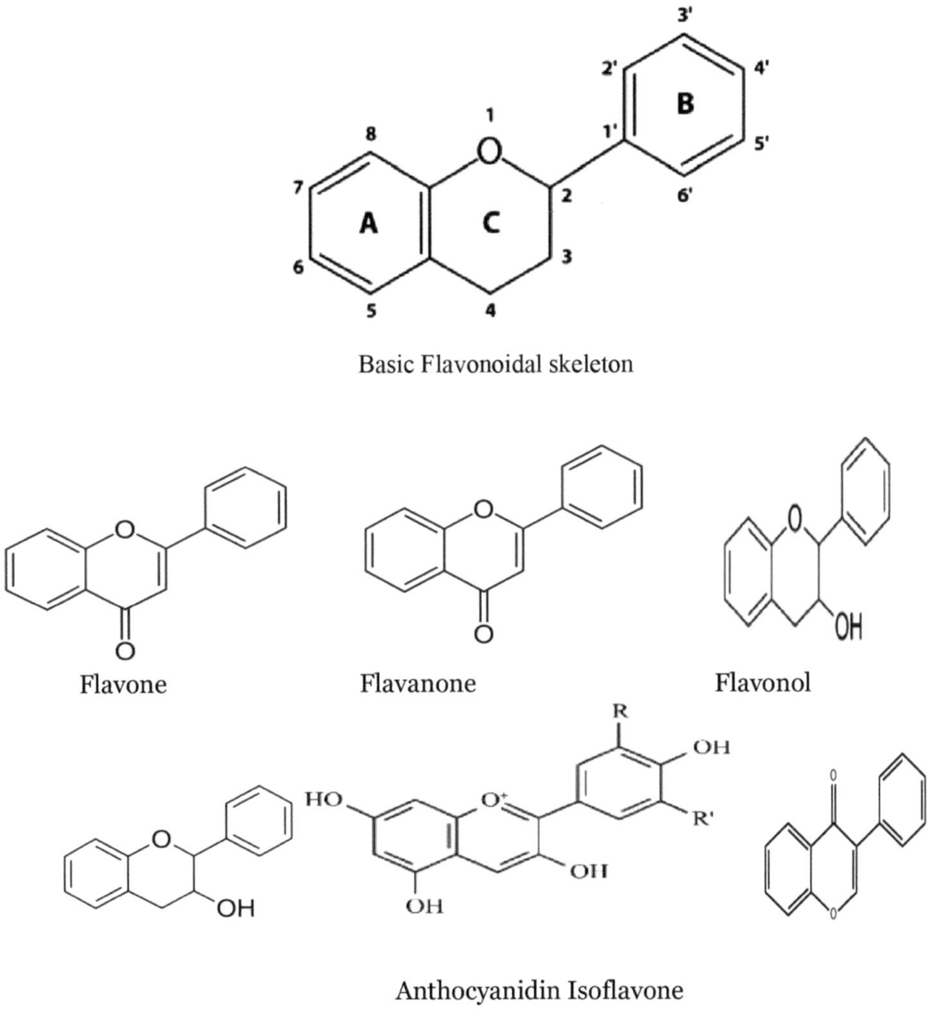

Figure 1.
Subclasses of flavonoids.

Flavonoids are commonly found in plants as O- or C-glycosides. Sugar substituents are destined to a hydroxyl group of the aglycone, commonly at site 3 or 7, in O-glycosides, whereas saccharide groups are connected to a carbon of the aglycone typically at site C6 or C8 in C-glycosides. Rhamnose, glucose, galactose, and arabinose are the most common saccharides. Plant metabolism, defense, signaling, disease, and symbiosis all benefit from flavonoids [3, 4]. These chemicals are accountable for floral color and are implicated in stress retort mechanisms, such as UV-B radiation [5, 6], microbial infection [7], and herbivore attacks by animals and insects [8].

2. Role of flavonoids in plant development

Polyphenols are a wide class of phenolic chemicals that includes flavonoids. Phenolic chemicals were important during evolution because they helped plants for the adaptation to life on land. At least 6000 molecules make up the flavonoid family, which could be split into, aurones, phlobaphenes, isoflavonoids, flavonols, flavones, anthocyanins, and flavonols [9]. The general phenylpropanoid route is utilized to produce these compounds from the starting compound phenylalanine. Several divisions of the general phenylpropanoid pathway supply precursors for synthesizing hundreds of chemicals. Lignins are structural polymers that give the secondary cell wall strength and stiffness and are necessary for waterproofing vascular cells [10]. Anthocyanin

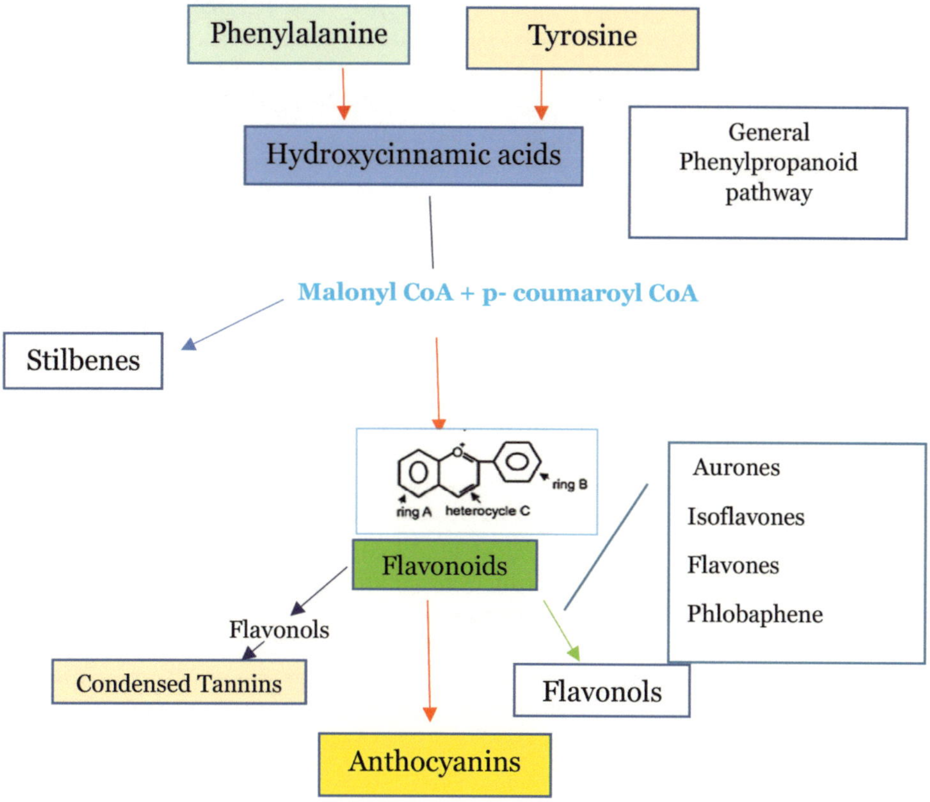

Figure 2.
Biosynthetic route for phenolic compounds.

pigments are generated from the flavylium cation (2-phenyl benzo pyrylium) and are glycosylated anthocyanidin precursors, as shown below in **Figure 2**.

This subgroup contains at least 400 molecules that range in hue dependent on pH, co-pigmentation, obtainable positive metallic ion, and backbone alterations. The color ranges from orange-red to purple. They are synthesized in the ground plasm and subsequently stored in the follicle. They are also present in cell membranes, chloroplasts, centers, and even the extracellular space, depending on the plant types, material, or phase of progress. Flavonoids, and phenolic chemicals, play a role in biotic stress resistance [11]. Since most of these chemicals have antibacterial and pesticide capabilities, serving as a vile and preventing pest progress and change, they may be constitutively produced or gather in retort to the bacterial incursion. In addition to their many activities in plants, Flavonoids have a wide range of medical, pharmacological, and nutritional qualities, earning them the moniker "nutraceutical" chemicals [12]. These metabolites provide promise for the prevention of a variety of illnesses, including cancer. They cause cancer cells to die, stimulate DNA repair, protect them from oxidative stress, and prevent cancer cells from multiplying [13].

3. Metabolism of flavonoids

In the uptake of flavonoids, two major compartments must be considered. The first compartment contains tissues such as the small intestine, liver, and kidneys. The colon is the body's second compartment (**Figure 3**). Flavonoids that have been consumed and then released with bile will make their way to the colon. Although around 40% of the absorbed ()-catechin was released in rats, the role of biliary secretion in humans is unclear in the small intestine with bile [14]. Metabolism of flavonoids in tissues and in the colon is discussed below in detail.

3.1 Metabolism in tissues

Biotransformation enzymes operate on flavonoids in the first compartment, including the small intestine and liver. Flavonol biotransformation enzymes can also be found in the kidney. Flavonoids and their colonic metabolites have been found to have polar hydroxyl groups conjugated with sulfate, glucuronic acid, or glycine [15]. Furthermore, O-methylation of flavonoids and their colonic metabolites by the enzyme catechol-O-methyltransferase is significant in the inactivation of the catechol moiety, that is, the two contiguous (ortho) aromatic hydroxyl groups. The conjugation reactions are exceedingly efficient in humans, as indicated by the fact that flavonoids primarily appear as conjugates in plasma and urine and that flavonoid aglycones in plasma are difficult to detect since they are mainly below the analytical methods' detection limits. Differential (HPLC) tests demonstrate the existence of flavonoid conjugates in humans, including O-methylated couples, with and deprived of hydrolysis of the model with a combination of b-glucuronidases and sulfatases: flavonols, flavones, catechins, flavanones, and anthocyanins. Anthocyanins, on the other hand, take a diverse approach. The indication is mounting that anthocyanidin glycosides can tolerate deglycosylation events in humans, at least in part. LC-MS [16] has revealed the presence of peonidin-3-glucoside, as well as peonidin-3-sambubioside [17] and pelargonidin-3-glucoside in urine.

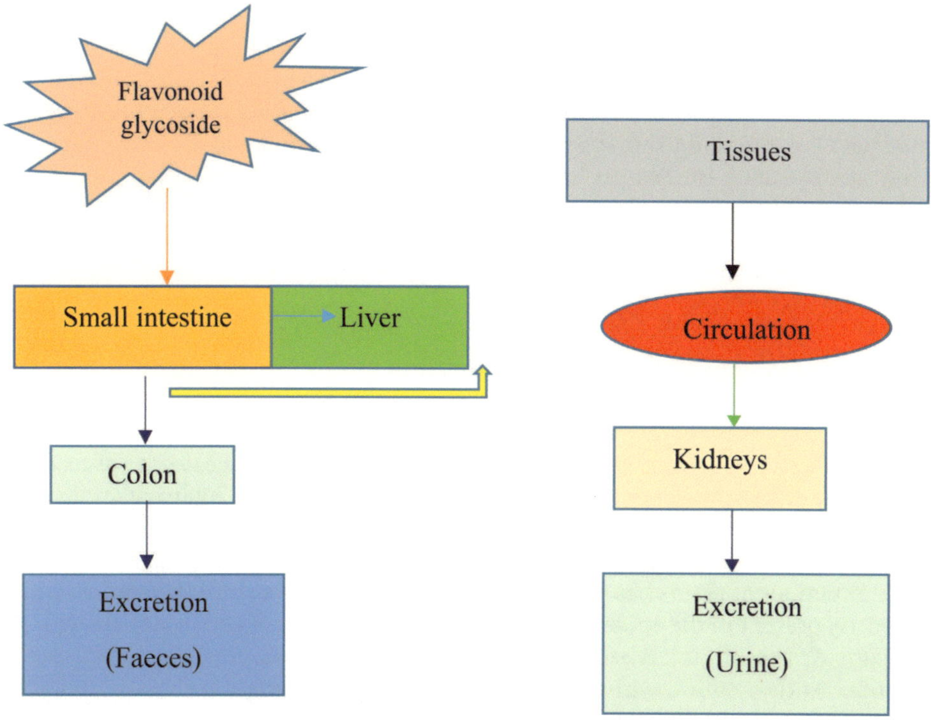

Figure 3.
Compartments involved in the metabolism of plant phenols.

3.2 Metabolism in the colon

Microbes break down the flavonoid fragment, splitting the flavonoid core, i.e., heterocyclic oxygen-containing ring, and the breakdown products are detected in urine and plasma. Several hydroxylated phenyl carboxylic acids are among them. Flavonols are broken down into phenylacetic acids and phenyl propionic acids. However, the effects of these phenyl propionic acids have yet to be proven in humans [18, 19]. In bodily tissues, these phenyl carboxylic acids are extra degraded by bacteria and transformed by enzymes. The phenyl propionic acids will be oxidized to benzoic acids as a result. Although roughly 60 possible phenolic acid metabolites were recognized and measured, only a small amount of phenolic acids were discovered. Hippuric acid, the glycine ester of benzoic acid, was an actual significant metabolite in people who had subsequent tea consumption [19, 20]. Microorganisms in the colon have been found to show an imperative role in the conversion of flavonoids to phenolic acids. Colonic bacteria generate glucuronidases, glycosidases, and sulfatases, which can shred flavonoid conjugates of their sugar moieties, glucuronic acids, and sulfates, in addition to the destruction of the flavonoid ring structure O-glycosides and C-glycosides that could be hydrolyzed by human gut bacteria [21].

3.3 The extent of metabolism of flavonoids

Flavonols were the initial to be examined, and human urine excretion was found to be quite modest. Only 0.1 percent to 3.6% of quercetin in the diet was eliminated in urine as quercetin conjugates. The sum of complete quercetin in blood plasma will be correlated with urinary excretion. Because the fascination of quercetin glucosides is high (up to 50%), while the evacuation of whole quercetin in urine is low, this indicates that quercetin undergoes substantial metabolism. Additional subclasses were investigated with helpers, and urine evacuation of metabolites with intact flavonoid structures was measured. Isoflavones have the highest rate of excretion of all flavonoids. Although isoflavones have a high bioavailability, flavonols glucosides have a higher bioavailability but a lower urine elimination. This would suggest that with isoflavones, the ring arrangement's metabolic modification level is less than through flavonols.

4. Recent advances in the role of flavonoids

Plant metabolites containing one oxygenated ring and two aromatic rings are known as flavonoids. Flavonoids are categorized depending on the degree of oxidation of their carbon rings; they can then undergo glycosylations, hydroxylations, acylations, methylations, or prenylations to modify their properties. As a result of these modifications, a new emergence of a huge number of different chemicals with varied functions in plants has occurred. UV wavelengths absorb all flavonoids, which are mostly present in the epidermis of plant cells and are produced in response to UV exposure. As a result, it has been proposed that they shield plants from this type of radiation. Anthocyanins, which absorb light in the visible range, are an example of flavonoids that absorb light at various wavelengths. Furthermore, certain flavonoids have antioxidant properties, which implies they serve as reactive oxygen species scavengers. However, most findings to date have remained based on in vitro studies, with little indication of how their functions are carried out in real life. In this assessment, we discuss recent advances in the study of the role of flavonols, flavones, and anthocyanins, three of the most prevalent flavonoids, in protecting plants from UV and high light exposure [22].

4.1 Advances in topical drug delivery

Topical delivery is one of the greatest popular methods for overcoming the disadvantages of other methods such as parenteral, oral, and so on. Oral distribution of phytochemicals is undesired due to the drug's distinctive flavor and odor, as well as the possibility of gastrointestinal (GI) breakdown until absorption [23].

4.1.1 Flavonols

Flavonols are O-glycosidic, ketonic chemicals with a 3-position sugar fraction. Flavonols are antioxidants that prevent the development of ROS. Due to UV, ozone radiation, and other damaging substances, the skin is the most prevalent target for oxidative stress. The combination of conjugated double bonds in the C-ring and adjacent hydroxyl groups in the B-ring gives flavonols their antioxidant properties [24]. Quercetin, kaempferol, myricetin, and other compounds are included in this class.

4.1.2 Quercetin

Quercetin is a flavonol in various foods, including leafy greens, citrus fruits, berries, and other fruits and vegetables. Quercetin inhibits edema, leukocyte production, and irritation. It helps reconstruct the skin's structure by endorsing the synthesis of new collagen fibers and generating ground substances [25]. Inflammatory mediators such as interleukins (IL) and prostaglandins (PGs), which are generated by COX, LOX, and LPS, are likewise suppressed by quercetin. The enzyme nitric oxide synthase, which also creates reactive nitrogen species like peroxynitrite, produces nitric oxide (NO), an inflammatory mediator. Quercetin is an antioxidant that prevents oxidative stress by inhibiting all of the mediators that cause it [26]. It avoids cell death by inhibiting the caspase-3 pathway and lowers mast cell development by inhibiting mast cell synthesis has an anti-allergic impact on histidine decarboxylase, IL-6, and monocyte chemoattractant protein (MCP-1).

4.1.3 Kaempferol

Kaempferol is primarily present in berries and plants of the allium and brassica family. It has antineoplastic, anti-inflammatory, and anti-allergic properties. It works as an anti-inflammatory drug by inhibiting NO synthase, which produces NO, a pro-inflammatory mediator. With the help of nuclear factor-inducing kinase (NIK) and mitogen-activated protein kinase (MAPKs), it also suppresses (NF-kappa B) [27]. It also inhibits COX-2 via decreasing nitric oxide synthase and TNF-, resulting in anti-inflammatory action. However, because kaempferol suffers significant first-pass metabolism and has a bioavailability of about 2%, topical application is preferable. UVB-induced cancer and photo-inflammation are treated with kaempferol, a new drug. It has been tested in skin cancer patients who have high levels of COX 2 enzymes. JB6 P+ mouse epidermal cells reduce AP-1 (Activator protein) activity via reducing COX-2. To test for AP-1 transactivation, JB6 P+ mouse epidermal cells were transfected with a luciferase reporter plasmid containing AP-1, and it was discovered that kaempferol reduces COX and AP-1 activities in a dose-dependent way, assisting in anticancer activity. The protooncogene tyrosine protein-kinase Src (Src) is a protooncogene that plays a key role in cell propagation, differentiation, and survival. Src activity is inhibited by kaempferol, which strives with ATP for the Src requisite position.

4.2 Flavanones

Consequent to the flavones origin, these fragrant ketones in citrus fruits like oranges and lemons. As a byproduct of citrus farming, a substantial quantity of hesperidin is produced. They are cytotoxic and inhibit tumor development; therefore, they might be used as anticancer drugs. Flavanones also serve as an anti-inflammatory and constrain protein tyrosine kinase, affecting cell development, distinction, mitosis, and death [28].

4.2.1 Hesperetin and hesperidin

Hesperidin has anti-inflammatory, antidiabetic, neuroprotective, and other properties. The aglycone component of hesperidin has antioxidant and anti-inflammatory properties by intruding with arachidonic acid, inhibiting COX and LOX enzymes, and limiting inflammatory mediator production [29]. It can whiten skin by reducing hyperpigmentation caused by UV radiation. Hesperetin and hesperidin generate anti-allergic action by constraining the issue of histamine from mast cells. They can also lower HMG CoA reductase and acyl CoA levels, resulting in a hypolipidemic effect. Hesperetin has a log P value ranging from 1.7 to 2.20, making it lipophilic and hard to absorb orally.

4.2.2 Naringenin

Naringenin has numerous properties, including antioxidant, antidiabetic, anti-inflammatory, and antineoplastic properties. It demonstrates antioxidants. MC1R: Melanocortin 1 receptor, MSH: Melanocyte stimulating hormone, MiTF: Microphthalmia associated transcription factor, Type-1: Tyrosinase related protein 1, Type-2: Tyrosinase related protein 2, ASIP: Agouti signaling protein] [– MSH: Melanocyte stimulating hormone, MC1R: Melanocortin 1 receptor, MiTF: Microphthalmia associated transcription factor, Type-2: Tyrosina Journal of Controlled Release 296 (2019) 190–201 R.L. Nagula, S. Wairkar 195 acts via chelating metal ions and constraining xanthine oxidase, averting oxygen radical generation and lipid peroxidation. It can also scavenge ROS through the −OH substitution. Because naringenin is hydrophobic and has low solubility and bioavailability, it can be used topically to create an effective formulation [30].

4.3 Flavanols

They are not to be mistaken with flavonol since they lack the ketone group. Epigallocatechin-3-gallate, proanthocyanidins, and other members of this class are included. Tea, chocolate, and a variety of vegetables and fruits contain them naturally [31]. The FDA, for use in a variety of medicinal formulations, has approved catechins and their derivatives.

4.3.1 Catechins

Catechins have antioxidant, photoprotective, anti-aging, anti-inflammatory, anticancer, neuroprotective, cardioprotective, antiviral, and antibacterial properties, among others. Epigallocatechin and epicatechin are abundant in grape seed extract and tea polyphenols. By scavenging free radicals, it has an antioxidant action [32]. Anti-inflammatory action is induced via inhibiting the COX enzyme, NO, PGs, and H2O2 production. Catechins help wounds heal faster by foraging free radicals at the wound site.

4.3.2 Anthocyanins

Anthocyanins are colorful glycosylated, water-soluble pigments that give fruits and vegetables their blue, red, and purple hues. They have been revealed to have antioxidant, anti-inflammatory, and depigmentation qualities in numerous scientific research.

Delphinidin contains anti-inflammatory, antioxidant, antitumorigenic, and antiangiogenic effects and is identified to suppress osteoclastogenesis in osteoporosis. Delphinidin inhibits the production of seditious intermediaries such as iNOS, NO, IL-6, MCP-1, and TNF-, which are produced when LPS inhibits the NF-B pathway and MEK1/2-ERK1/2 signaling [33]. It was tested for psoriasis on flaky skin mice and shown to reduce epidermal thickness. Infiltrating macrophages and caspase 14 downregulation were also seen. Keratin-14, which induces hyperproliferation, was also reduced. By inhibiting keratin-14, delphinidin serves as an antiproliferative. The level of pathological indicators of psoriasis lesions is reduced when delphinidin is applied to flaky mouse skin.

4.3.3 Role of anthocyanins in promoting human health

Anthocyanins may play a beneficial role in human health, according to numerous research. They function as neuroprotective agents and have antidiabetic and antiobesity properties. These substances could be helpful in lowering inflammation and protecting the heart [34]. Additionally, they appear to be effective in halting and preventing the spread of cancer. The biological activity of anthocyanins in rats was recently confirmed by a study by Vanzo and colleagues [35], where the ability of anthocyanins to influence mammalian metabolism was shown in an investigation of metabolomic changes in the brain and plasma of adult rats after intravenously administering cy-3- glc [36]. In the blood, kidneys, and liver of rats, it was demonstrated that cy-3-glc changes a number of significant cellular metabolites, including bile acids, glutathione, oxidized glutathione, and certain lipids [35]. Due to the high anthocyanin concentration in blueberries, this fruit may be a food that improves or promotes health. Routray and Orsat [37] provided evidence for this in a study that analyzed a number of factors related to the potential health effects of anthocyanins, emphasizing understanding [35].

Prebiotics present in brown rice, such as arabinoxylan and -glucan, are advantageous for the Bifidobacterium and Lactobacillus that make up the human gut microbiota. They are thought to play a part in creating an anti-obesity impact. Additionally, brown rice was employed as a preventative measure for type 2 diabetes due to its antidiabetic benefits. This is probably because one of their constituents, −oryzanol, is crucial in regulating the ER stress brought on by a high-fat diet in the hypothalamus, which aids in lowering the desire for fatty foods. Additionally, brown rice's oryzanol has been shown to lower blood cholesterol levels and stop pancreatic cells from dying. Through their antioxidant action, dietary rice brans, which give brown rice its brown color, also demonstrate powerful anticancer properties [37, 38].

5. Biotechnological applications of flavonoids

Accepting the complex control of flavonoid production has apparent implications, such as the generation of distinct flower colors and fruit types with appealing esthetic and/or agronomic traits, thereby increasing natural selection that has happened from the beginning of time. Petunia an1 (bHLH) or an2 (MYB) variations, morning glory Ipivs (bHLH), c (InMYB1), and ca (InWDR1) mutants, and gentian GtMYB3 mutants have all been characterized as having flowers with a diversity of coloration produced by mutations in the encrypting arrangement of one or more components of the MBW composite. An alteration produces the lack of coloration in fruits in the coding arrangement of the MYB genetic factor VvMYBA2 and MrMYB1, as well as a jumping gene pullout in the promoter of the MYBA1 gene. Grape berry and Chinese bayberry are two examples. Manipulation of flavonoid production to produce fruit and vegetables high in antioxidants and nutritious components, befitting the moniker "superfruit. Nutraceuticals," would be of prodigious importance to human wellbeing than ever [39]. Anthocyanin accumulation was caused by the ectopic appearance of the MYB-encoding gene LeANT1 in tomato skin and subepidermal cell layers. Likewise, co-expression of the bHLH Delila and MYB Rosea 1 genes below the regulator of the fruit-specific promoter E8 resulted in a significant upsurge in anthocyanin pigments in the flesh and skin resulting in dark purple fruits. Their lifespan was significantly increased when cancer-prone p53 knockout mice were given these transgenic tomatoes. This study is the first step in developing fruits that are high in flavonoid bioactive components and might be part of a healthy daily diet. MdMYB10 and IbMYB1 are two genes that control anthocyanin accumulation in apples [40]. This flavonoid is found by modifying the expression of these commonly eaten foods. The content of these foods might be raised. In tomatoes, constitutive expression of ZmLc, Delila, and MYC-RP/GP led to anthocyanin accumulation in aerial tissues and roots, suggesting that plant transformation with bHLH transcription factors may be investigated. Finally, increasing PA content in forage crops (mainly alfalfa and clover) may assist in preventing pasture bloat in ruminant animals by delaying fermentation in the rumen. PA accumulation may arise from overexpression of ZmLc in alfalfa leaves. Overexpression of ZmSn in the bird's foot trefoil increased PA biogenesis and anthocyanin accumulation in certain leaf areas. However, constitutive expression of a transgene in many circumstances under Ecological pressure, such as cold and bright light, is necessary since a heterologous system is insufficient to stimulate flavonoid accumulation automatically. In Arabidopsis 35S::PAP1 plants, for example, poor growth circumstances led to the downregulation of positive regulators and the overexpression of putative transcriptional repressors AtMYB6, AtMYB3, and AtMYBL2 [41].

6. Utilizing modern technology to research the flavonoid pathway

We currently have extraordinary knowledge about how different chemical components in plants are controlled in abundance, thanks to the recent rapid development of metabolomics and the use of varied populations for genetic mapping [42]. In a population of rice Zhenshan 97 and Minghui 63 recombinant inbred lines (RILs), metabolic QTL were discovered using high-throughput genotyping and metabolomics data, and some of the candidate genes for flavonoid content were further validated by looking at over-expression transgenic rice lines [43]. Flavonol 3-O-gentiobioside 7-O-rhamnoside (F3GG7R) synthesis in an Arabidopsis RIL population has recently been linked to a novel gene (BETA GLUCOSIDASE 6; BGLU6) [44, 45]. Following genome wide association studies (GWAS) on a diverse maize population that revealed the genetic effects underpinning metabolic heterogeneity, hundreds of loci related with metabolites from numerous pathways, including flavonoid metabolism, were found in maize [46]. The co-expression and direct target genes of the R2R3-MYB transcription factor P1 were also studied using near isogenic lines (NILs) carrying P1-rr and P1-ww. This discovery marked a significant advancement in our understanding of P1's gene regulation circuitry because targeted molecular tests showed that P1 regulates some well-known genes involved in flavonoid biosynthesis, such as FLS1 and A1 [47]. The corn earworm (*Helicoverpa zea*), which may cause significant damage to maize in the Americas, is naturally resistant to maysin (C-glycosyl flavone), which is contained in maize silks. Through QTL mapping [29] in 2004, two loci that can impart the salmon silks phenotypes salmon silks 1 (sm1) and salmon silks 2 (sm2) were found. Additionally, earlier genetic investigations suggested that P1 would be epistatic to the salmon silk mutation [13]. The molecular identification of the sm1 and sm2 gene products is revealed as an UDP-rhamnose synthase and a rhamnosyl transferase, respectively, based on the knowledge of the genes regulated by P1 and the existing sm1 and sm2 mapping information [48, 49]. The maysin biosynthetic pathway is therefore finished with the molecular characterization of sm1 and sm2.It can thus be anticipated that deep probing of further profiling studies will facilitate the elucidation of the genetic complexity of maize flavonoid biosynthesis. Indeed, integrative approaches are increasingly applied to enhance our understanding of metabolic pathway structure and regulation and how these affect the end-phenotypes of plants [50].

7. Conclusion

Over the last 5 years, our understanding of the metabolism of entire subclasses of flavonoids has progressively improved. Flavonoids are probable nutraceuticals abundantly distributed in vegetables and fruits, given the special focus on wellbeing and illness anticipation over stable nourishment incorporating ordinary goods. This information is crucial for fully assessing their possible health implications, and it still needs to be expanded. We still need to explore information on the quantities and metabolic forms of flavonoids that tissues and cells get exposed to after their consumption. The next stage is investigating possible biological impacts at the tissue and cellular levels. New genomic approaches will open up a world of possibilities in this discipline. Knowing which metabolites will reach tissues and cells, at what concentrations, and to what extent they will be taken up and changed in cells after a flavonoid-rich diet is crucial. The high-throughput genomics technologies will then help us better understand how flavonoids influence metabolic paths and, as a result, improve social health.

Acknowledgements

The authors are thankful to the Institute of the Pharmaceutical Technology, Sri Padmavati Mahila Visvavidyalayam.

Conflict of interest

The authors declare that they have no competing interests to declare.

Abbreviations

DNA	deoxyribonucleic acid
HPLC	High Performance Liquid Chromatography
LC-MS	Liquid Chromatography-Mass Spectrometry
UV	ultra violet
GI	gastro intestinal
ROS	reactive oxygen species
IL	interleukins
PGs	prostaglandins
COX	cyclooxygenase
LOX	lipoxygenase
LPS	lipopolysaccharide
NO	nitric oxide
MCP	monocyte chemoattractant protein
NIK	nuclear factor-inducing kinase

MAPK	mitogen-activated protein kinase
PAs	proanthocyanidins
TNF	tumor necrosis factor
AP	activator protein
Src proto	oncogene tyrosine-protein kinase
ATP	adenosine tri phosphate
HMG-CoA β	hydroxy β-methylglutaryl-CoA
MC1R	melanocortin 1 receptor
MSH	melanocyte stimulating hormone
MiTF	microphthalmia associated transcription factor
ASIP	agouti signaling protein
MiTF	microphthalmia associated transcription factor
FDA	Food and Drug Administration
MEK	mitogen-activated protein kinase
ERK	extracellular signal-regulated kinase
TFs	transcription factors
BGLU6	beta glucosidase 6
F3G	flavonol 3-O-glucoside
FLS1	flavonoid 7-O-glucoside; FLS1
BGLU6	beta glucosidase
F3GG7R	flavonol 3-O-gentiobioside 7-O-rhamnoside
NILs	near isogenic lines
sm1	Salmon silks 1
sm2	Salmon silks 2
GWAS	genome wide association studies

Author details

Indireddy Theja[1] and Banoth Ramya Kuber[2]*

1 Department of Pharmaceutical Analysis, Institute of Pharmaceutical Technology, Sri Padmavati Mahila Visvavidyalayam (Women's University), Tirupati, Andhra Pradesh, India

2 Department of Pharmacognosy, Institute of Pharmaceutical Technology, Sri Padmavati Mahila Visvavidyalayam (Women's University), Tirupati, Andhra Pradesh, India

*Address all correspondence to: rkuberpharma@yahoo.com

References

[1] Hodek P, Trefil P, Stiborová M. Flavonoids, potent and versatile biologically active compounds interacting with cytochromes P450. ChemicoBiological Interactions. 2002;**139**:1-21

[2] Marais JPJ, Deavours B, Dixon RA, Ferreira D. The stereochemistry of flavonoids. In: Grotewold E, editor. The Science of Flavonoids. New York: Springer Science Business Media; 2006. pp. 1-46

[3] Gould KS, Lister C. Flavonoid functions in plants. In: Andersen RM, Markham KR, editors. Flavonoids: Chemistry, Biochemistry, and Applications. New York: Taylor & Francis Group, CRC Press; 2006. pp. 397-441

[4] Bravo L. Polyphenols: Chemistry, dietary sources, metabolism, and nutritional significance. Nutrition Reviews. 1998;**56**:317-333

[5] Rosenberg Zand RS, Jenkins DJA, Diamandis EP. Flavonoids and steroid hormone-dependent cancers. Journal of Chromatography B. 2002;**777**:219-232

[6] Olsson LC, Veit M, Weissenböck G, Bornman JF. Differential flavonoid response to enhanced UV-B radiation in Brassica napus. Phytochemistry. 1998;**49**:1021-1028

[7] Middleton EM, Teramura AH. The role of flavonol glycosides and carotenoids in protecting soybean from ultraviolet-B damage. Plant Physiology. 1993;**103**:741-752

[8] Izaguirre MM, Mazza CA, Biondini M, Baldwin IT, Ballare CL. Remote sensing of future competitors: Impacts on plant defenses. Proceedings of the National Academy of Sciences. 2006;**103**:7170-7174

[9] Harborne JB, Williams CA. Advances in flavonoid research since. Phytochemistry. 1992;**55**:481-504

[10] Vanholme R, Demedts B, Morreel K, Ralph J, Boerjan W. Lignin biosynthesis and structure. Plant Physiology;**153**:895-905

[11] Bhattacharya A, Sood P, Citovsky V. The roles of plant phenolics in defence and communication during Agrobacterium and Rhizobium infection. Molecular Plant Pathology. 2010;**11**:705-719

[12] Lin JK, Weng MS. Flavonoids as nutraceuticals. In: Grotewold E, ed. The Science of Flavonoids. Berlin: Springer; pp. 213-238

[13] Khan N, Adhami VM, Mukhtar H, Khan N, Adhami VM, Mukhtar H. Apoptosis by dietary agents for prevention and treatment of prostate cancer. Endocrine-Related Cancer. 2011;**17**:39-52

[14] Cody V, Middleton E, Harborne J, . Plant Flavonoids in Biology and Medicine Biochemical, Pharmacological, Structure-Activity Relationships. New York: Alan R. Liss; 1982. pp. 177-194

[15] Hollman PCH, Katan MB. Absorption, metabolism, and bioavailability of flavonoids. In: Rice-Evans C, Packer L, editors. Flavonoids in Health & Disease. New York: Marcel Dekker; 1998. pp. 483-522

[16] Felgines C, Talave'ra S, Gonthier M-P, Texier O, Scalbert A, Lamaison J-L, et al. Strawberry anthocyanins are recovered in urine as glucuro- and sulfoconjugates

in humans. The Journal of Nutrition. 2003;**133**:1296-1301

[17] Wu X, Cao G, Prior RL. Absorption and metabolism of anthocyanins in elderly women after consumption of elderberry or blueberry. The Journal of Nutrition. 2002;**132**:1865-1871

[18] Rechner AR, Kuhnle G, Bremner P, Hubbard GP, Moore KP, Rice-Evans CA. The metabolic fate of dietary polyphenols in humans. Free Radical Biology & Medicine. 2002;**33**:220-235

[19] Olthof MR, Hollman PCH, Buijsman MNCP, van Amelsfoort JM, Katan MB. Chlorogenic acid, quercetin-3-rutinoside, and black tea phenols are extensively metabolized in humans. The Journal of Nutrition. 2003;**133**:1806-1814

[20] Olthof MR, Hollman PCH, Vree TB, Katan MB. Bioavailabilities of quercetin-3-glucoside and quercetin-40-glucoside do not differ in humans. The Journal of Nutrition. 2002;**130**:1200-1203

[21] Scheline RR. Metabolism of foreign compounds by gastrointestinal microorganisms. Pharmacological Reviews. 1973;**25**:451-523

[22] Shelke SJ, Shinkar DM, Saudagar RB. Topical gel: A novel approach for development of topical drug delivery system. International Journal of Pharmacy and Technology. 2013;**5**:2739-2763

[23] Zaki NM. Progress and problems in nutraceuticals delivery. Journal of Bioequivalence & Bioavailability. 2014;**6**:075-077

[24] Wang H, Helliwell K. Determination of favonols in green and black tea leaves and green tea infusions by. High-performance Liquid Chromatography. 2001;**34**:223-227

[25] Andrea GD. Fitoterapia quercetin: A flavonol with multifaceted therapeutic applications? Fitoterapia. 2015;**106**:256-271

[26] Park EJ, Kim JY, Jeong MS, Park KY, Lee MW, Joo SS, et al. Effect of topical application of quercetin-3-O-(2″-gallate)-α-l-rhamnopyranoside on atopic dermatitis in NC/Nga mice. Journal of Dermatological Science. 2015;**77**:166-172

[27] Chen AY, Chen YC. A review of the dietary flavonoid. Kaempferol on Human Health and Cancer Chemoprevention. 2014;**138**:2099-2107

[28] Shi L, Feng XE, Cui JR, Fang LH, Du GH, Li QS. Synthesis and biological activity of flavanone derivatives. Bioorganic & Medicinal Chemistry Letters. 2010;**20**:5466-5468

[29] Cho J. Antioxidant and neuroprotective effects of hesperidin and its aglycone hesperetin. Archives of Pharmacal Research. 2006;**29**:699-706

[30] Joshi R, Kulkarni YA, Wairkar S. Pharmacokinetic, pharmacodynamic and formulations aspects of Naringenin: An update. Life Sciences. 2018;**215**:43-56

[31] Grzesik M, Bartosz G, Dziedzic A, Narog D, Namiesnik J, Sadowska-Bartosz I. Antioxidant properties of ferrous flavanol mixtures. Food Chemistry. 2018;**268**:567-576

[32] Grzesik M, Naparło K, Bartosz G, Sadowska-Bartosz I. Antioxidant properties of catechins: Comparison with other antioxidants. Food Chemistry. 2018;**241**:480-492

[33] Sobiepanek A, Milner-Krawczyk M, Bobecka-Wesołowska K, Kobiela T. The effect of delphinidin on the mechanical properties of keratinocytes exposed to UVB radiation. Journal of

Photochemistry and Photobiology B: Biology. 2016;**164**:264-270

[34] Quattrocchio F, Wing J, van der Woude K, Souer E, de Vetten N, Mol J, et al. Molecular analysis of the anthocyanin2 gene of petunia and its role in the evolution of flower color. The Plant Cell. 1999;**11**:1433-1444

[35] Vanzo A, Scholz M, Gasperotti M, Tramer F, Passamonti S, Vrhovsek U, et al. Metabonomic investigation of rat tissues following intravenous administration of cyanidin 3-glucoside at a physiologically relevant dose. Metabolomics. 2013;**9**:88-100

[36] Pojer E, Mattivi F, Johnson D, Creina S. Stockley the case for anthocyanin consumption to promote human health: A review. Comprehensive Reviews in Food Science and Food Safety. 2013;**12**:483-508

[37] Routray W, Orsat V. Blueberries and their anthocyanins: Factor affecting biosynthesis and properties. Comprehensive Reviews in Food Science and Food Safety. 2011;**10**:303-320

[38] Kennedy G, Burlingame B. Analysis of food composition data on rice from a plant genetic resources perspective. Food Chemistry. 2003;**80**(4):589-596

[39] Kobayashi S, Ishimaru M, Hiraoka K, Honda C. Myb-related genes of the Kyoho grape (*Vitis labruscana*) regulate anthocyanin biosynthesis. Planta. 2002;**215**:924-933

[40] Mathews H, Clendennen SK, Caldwell CG. Activation tagging in tomato identifies a transcriptional regulator of anthocyanin biosynthesis, modification, and transport. The Plant Cell. 2003;**15**:1689-1703

[41] Rowan DD, Cao M, Lin-Wang K. Environmental regulation of leaf colour in red 35S: PAP1 Arabidopsis thaliana. New Phytologist. 2009;**182**:102-115

[42] Wen W, Brotman Y, Willmitzer L, Yan J, Fernie AR. Broadening our portfolio in the genetic improvement of maize chemical composition. Trends in Genetics. 2016. DOI: 10.1016/j.tig.2016.05.003

[43] Gong L, Chen W, Gao Y, Liu X, Zhang H, Xu C, et al. Genetic analysis of the metabolome exemplified using a rice population. Proceedings of the National Academy of Sciences of the United States of America. 2013;**110**(50):20320-20325

[44] Ishihara H, Tohge T, Viehöver P, Fernie AR, Weisshaar B, Stracke R. Natural variation in flavonol accumulation in Arabidopsis is determined by the flavonol glucosyltransferase BGLU6. Journal of Experimental Botany. 2015;**67**(5):1505-1517

[45] Schaffner AR. Flavonoid biosynthesis and Arabidopsis genetics: More good music. Journal of Experimental Botany. 2016;**67**(5):1203-1204

[46] Wen W, Li D, Li X, Gao Y, Li W, Li H, et al. Metabolome-based genome-wide association study of maize kernel leads to novel biochemical insights. Nature Communications. 2014;**5**:3438

[47] Falcone Ferreyra ML, Rius S, Emiliani J, Pourcel L, Feller A, Morohashi K, et al. Cloning and characterization of a UV-B-inducible maize flavonol synthase. The Plant Journal. 2010;**62**(1):77-91

[48] Szalma SJ, Buckler ES, Snook ME, McMullen MD. Association analysis of candidate genes for maysin and chlorogenic acid accumulation in maize silks. Theoretical and Applied Genetics. 2005;**110**(7):1324-1333

[49] Morohashi K, Casas MI, Falcone Ferreyra ML, Mejia-Guerra MK, Pourcel L, Yilmaz A, et al. A genome-wide regulatory framework identifies maize pericarp color1 controlled genes. The Plant Cell. 2012;**24**(7):2745-2764

[50] Tohge T, Scossa F, Fernie AR. Integrative approaches to enhance understanding of plant metabolic pathway structure and regulation. The Plant Journal. 2015;**169**(3):1499-1511

Chapter 7

Flavonoids: Recent Advances and Applications in Crop Breeding

Shuchi Nagar, Saurabh Dey, Aishik Das and Soumya Basu

Abstract

Flavonoids are secondary metabolites that perform a wide range of roles in plants. These include their involvement in plant growth, pigmentation, and UV protection, to a variety of defense and signaling activities. Flavonoids such as chalcones, flavones, flavanols, anthocyanins, and proanthocyanins are widely distributed in the plant kingdom. The metabolic routes of the flavonoids are exploited extensively using several biotech approaches to enhance the crop variety and incorporate varied nutritional benefits. Many flavonoids are key components of medicinal plants and possess nutritional significance. Specific mutations in flavonoid-related genes are typically responsible for the diversity in flavonoids, resulting in quantitative and qualitative variations in metabolic profiles. Thereby numerous attempts have been made to increase flavonoid content in agronomically important species. Flavonoids are also employed in the regulation of inflammation, in arthritis, and in cancer prevention strategies, due to their ubiquity in the human diet. Advances in the comprehension of flavonoid biosynthesis and modulation have prompted a surge in researches aiming at modifying the flavonoid pathway to improve nutritional value, plant defenses against infections and the feeding value of livestock. This chapter briefly discusses the varied role of flavonoids, their biosynthesis, and their distribution over the plant kingdom. Furthermore, it exclusively highlights the several biotech-based trending pieces of research based on introducing flavonoid biosynthesis in commercial crops.

Keywords: biosynthesis, crop breeding, flavonoids, genetic engineering, mediterranean crops, plasticity

1. Introduction

1.1 What are flavonoids?

Flavonoids are naturally occurring secondary metabolites in plants, possessing a polyphenolic structure. They are widely distributed in the leaves, seeds, barks, flowers, fruits, and vegetables. Over 8000 flavonoids have been identified to date and most of them have been found to be engaged in various biological activities in plants, animals, and bacteria. Apart from imparting pigmentation in plants, flavonoids afford protection against UV radiation, herbivores, and pathogens [1–3]. In addition, flavonoids have also been found to serve as detoxifying and antimicrobial defense agents in the animal world. They appear to have played a significant part in the effectiveness of ancient medicinal therapies, and since then, their usage has continued to this day. Notably, detailed pieces of evidence from various studies have confirmed flavonoids' role in growth metabolism and gene regulation as well [4].

1.2 Structure and types of flavonoids

The name 'Flavonoids' refers to a group of plant pigments generated mostly from benzo-γ-pyrone (**Figure 1**). Advanced techniques such as [1H-1H]-correlated spectroscopy, [1H]- and [13C]-NMR spectrometry, X-ray diffraction, mass spectrometry, circular dichroism, and optical rotatory dispersion help us to analyze and elucidate the flavonoid structures and configurations [4]. The class of flavonoids primarily comprises of anthocyanidins, proanthocyanins, flavonols, iso-flavonoids, chromones, flavones, iso-flavones, flavanes, flavanones, flavanols, catechins, aurones, benzofurones, and coumarins.

The variability observed in the flavonoids mainly occurs due to differences in the following features:

1. Changes in the aglycone's ring structure and state of oxidation/reduction.

2. Variations in the aglycone's hydroxylation extent and the locations of hydroxyl groups.

3. Various methods of derivatizing the hydroxyl groups, such as using methyl groups, polysaccharides, or isoprenoids [4].

Flavonoids exhibit a range of beneficial effects in human health like

1. anticholinesterase activity and combating neurodegenerative diseases [5].

2. anti-inflammatory activity [6].

3. steroid-genesis modulators [5].

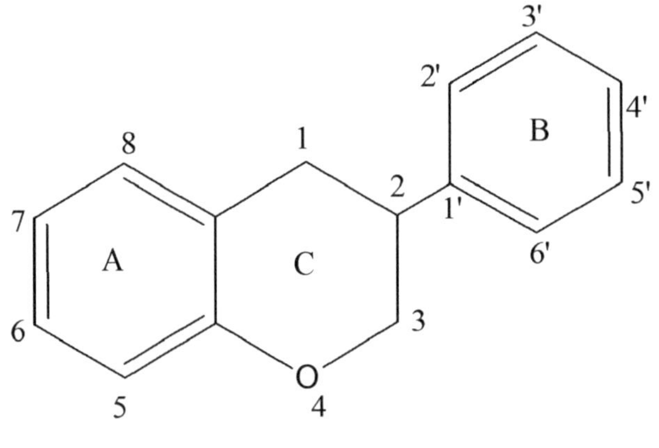

Figure 1.
Basic structure of flavonoid.

4. xanthine oxidase modulator [7].

5. radical scavenging agent [8].

6. anti-carcinogenic activity [6].

1.3 Role of flavonoids in plants

Flavonoids are responsible for distinct flavor and color, which draw pollinators, as well as characteristic color and fragrance of flowers. Additionally, they facilitate the germination of seeds and spores, as well as the growth and development of seedlings, by fruit dispersal. Plants can be protected from biotic and abiotic challenges by flavonoids, which also operate as UV filters, signal molecules, phytoalexins, detoxifying agents, and antimicrobial defense components. They are recognized for their ability to frequently play a useful role in the capacity of plants to adapt to heat, to cold, to frost, to drought, and to both. Early advances in floral genetics have mostly been made by mutation techniques that impact flower colors that are produced from flavonoids, and it has been proven that plants that are involved in flavonoid production are capable of functional gene silencing [1].

1.3.1 Role in pigmentation

Flavonoids play a prominent role in floral coloration, as well as pollinator attractiveness and UV protection [9, 10]. A study on Papaver flowers has revealed that the spatial occurrence of flavonoids is responsible for its wide-range variation in flower color [11]. Approximately 8000 flavonoids contribute to the vibrant colors seen in fruits, herbs, vegetables, and medicinal plants.

1.3.2 Role as a growth regulator

Flavonoids have the ability to regulate auxin movement and catabolism. Recent research has revealed that flavonoids are capable of modulating protein activity during cell growth [12].

1.3.3 Role in nitrogen metabolism

Flavonoids, through inducing root nodulation, play a significant role in nitrogen metabolism in nitrogen-fixing plants. Dinitrogen-fixing bacteria, such as the *Rhizobium* strain, exist in symbiosis with leguminous plants and are found in these nodules. The major action of flavonoids is likely to be the stimulation of genes that express proteins necessary by nodule cells, but because they are antioxidative, they are also well adapted to participate in dioxygen removal [4].

1.3.4 Role in combating oxidative stress

As a result of a variety of biotic and abiotic stimuli causing oxidative stress, reactive oxygen species (ROS) and reactive nitrogen species (RNS) are produced in plants. Nearly all of the increase in flavonoid synthesis in plants comes from oxidative stress. Flavonoids like mono-apigenin and mono-kaempferol and dihydroxy B-ring-substituted (Luteolin and Quercetin) flavonoid glucosides reduce ROS generation, quench ROS once they have formed, and absorb the UV-B and UV-A wavelengths. When early plants went from the water to the land, flavonoids performed key UV-B screening roles. The type of substitution on distinct rings of flavonoids determines the extent of antioxidant capability and ability to absorb UV wavelengths [12].

1.3.5 Role in defense against pathogens and insects

Flavonoids in plants are also helpful in protecting plants from harmful bacteria and fungus. Catechins and other flavanols potentially act as a plant's defensive mechanism against insects that are damaging to it [13–15].

2. Flavonoid metabolism and Biosynthesis

2.1 Distribution of Flavonoid subclasses in the plant kingdom

According to study, flavonoids may be found in angiosperms, gymnosperms, and pteridophytes. Due to the wealth of information available on flavonoids in many species, flavonoid subclasses (such as anthocyanins, chalcones, flavones, flavonols, and proanthocyanidins) are present in each subgroup of plants can be identified. Flavone and flavanone are present in all plant groups, with the exception of hornworts. Plant families that produce flavonoid subclasses have evolved and diversified as well. For instance, the angiosperms have the most varied flavonoid aglycones. The liverworts *Radula variabilis* and *Radula spp.* contain prenyldihydrochalcone, whereas more than 1000 prenylflavonoids have been found in legumes. These results indicate that either the two plant groups independently evolved the ability to make prenylflavonoids or that many species lost this capacity over evolution. Flavonoid molecules show that plants have genes for the manufacture of flavonoids. Therefore, analytical approaches for identifying flavonoids are necessary to comprehend the evolution of flavonoid metabolism in the plant kingdom [16] (**Figure 2**).

2.2 Evolution of Flavonoid metabolism

The enzymes, chalcone isomerase (CHI) and isoflavone reductase in Chlamydomonas, dihydrokaempferol-4-reductase and naringenin chalcone synthase (CHS) in *Phaeodactylum*, and CHI and dihydroflavonol reductase in *Ectocarpus* were created as a result of several evolutionary processes in representatives of bryophytes (mosses), liverworts, and hornworts. CHI-like enzymes were discovered in certain proteobacteria and fungi, and they may have been acquired by horizontal gene transfer. Contrarily, the recruitment and gene duplication of polyketide synthases and oxoglutarate-dependent dioxygenases from primary metabolism, respectively, led to the evolution of CHS and F3H. The first three flavonoids, chalcones, flavanols, and flavones, were created as a result of the CHS, CHI, and F3H activities. These metabolites, which have not altered in 500 million years, are essential intermediates in today's irreducibly complicated flavonoid manufacturing pathways in plants [17, 18].

The number of key events that sparked the flavonoid pathway's gradual rise, variety, and evolutionary successes are:

- the recruitment of enzymes from fundamental metabolisms, such as the polyketide, phenylpropanoid, and shikimate pathways

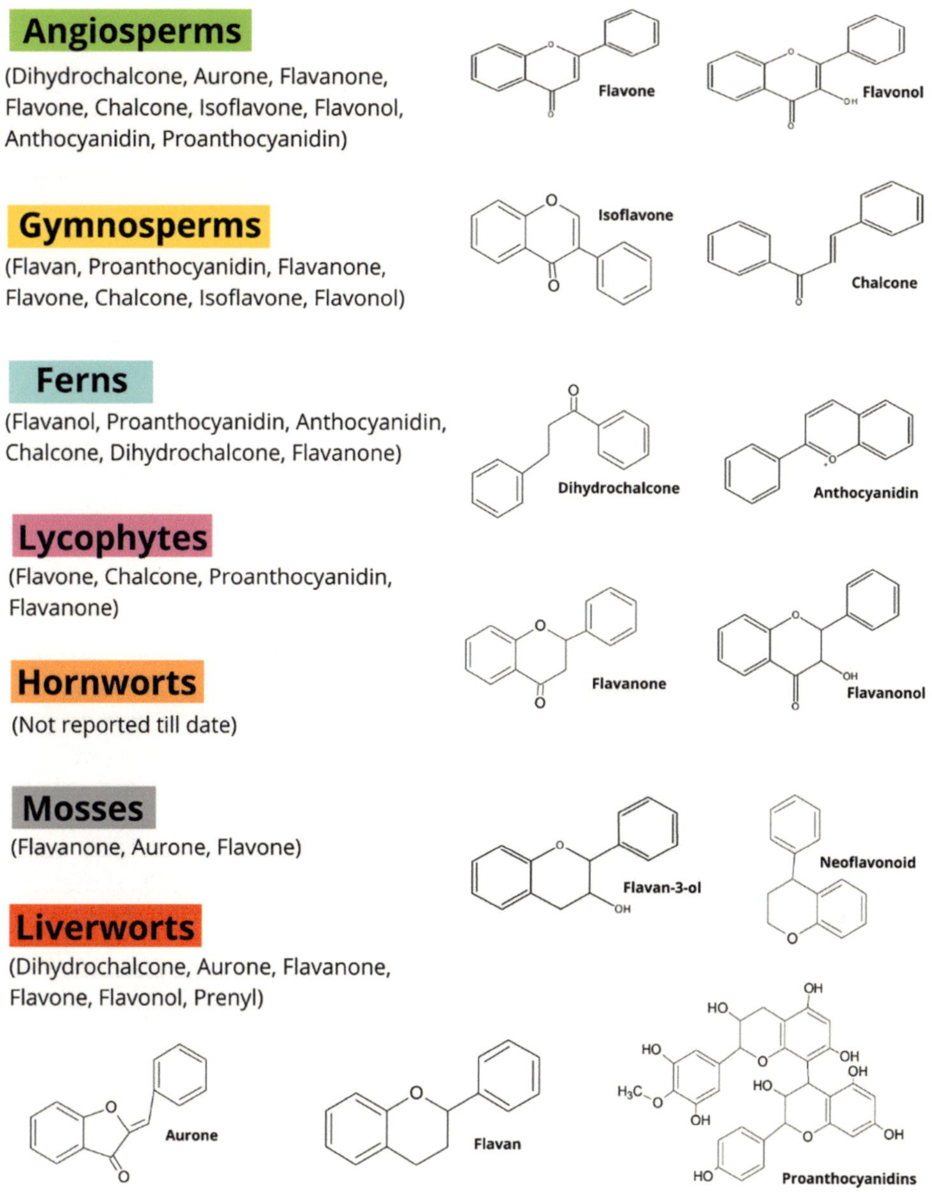

Figure 2.
Distribution of flavonoids in plant kingdom and their respective structures.

- horizontal gene transfer between bacteria and fungi in plant/algal symbioses
- variations in the substrate selectivity and regiospecificity (the ability to change specific regions of the substrate molecules) of metabolic enzymes (ability to bind to different substrates)
- modifications in the regulation of the flavonoid gene

- the flexibility of flavonoid pathways and their capacity to shift intermediate molecule fluxes towards the production of complex scaffolds of quite varied chemicals depending on the needs of the local ecosystem.

More than 10,000 of these compounds have been found in over 9000 current plant species as a result of these evolutionary processes, making flavonoids one of the most extensively distributed routes in plants today [16, 19].

2.3 The flavonoid biosynthetic pathways

From a genetic standpoint, a lot of work has been done to decipher the flavonoids' biosynthesis routes. Flavonoid synthesis mutants have been discovered in a variety of plant species. The first important experimental models in this system were snapdragon (*Antirrhinum majus*), petunia (*Petunia hybrida*), and maize (*Zea mays*), which led to further discovery of several structural and regulatory flavonoid genes. Arabidopsis (*Arabidopsis thaliana*) has recently contributed to the study of flavonoid pathway regulation and subcellular localization [20].

2.3.1 Following are the biosynthetic pathways of some major flavonoids

The Figure shows the eight branches of flavonoid biosynthetic pathway (showed in the eight different colored boxes) and four important intermediate metabolites (represented by the green boxes) (**Figure 3**). The abbreviated forms of enzyme names and flavonoid compounds mentioned in the figure are as follows: (i) ANR: anthocyanidin reductase; (ii) ACCase: acetyl-CoA carboxylase; (iii) AS: aureusidin synthase; (iv) 4CL: 4-coumarate: CoA ligase; (v) CHS: chalcone synthase; (vi) CHI: chalcone isomerase; (vii) CHR: chalcone reductase; (viii) C4H: cinnamic acid 4-hydroxylase; (ix) CH2′GT: chalcone 2′-glucosyltransferase; (x) CH4′GT: chalcone 4′-O-glucosyltransferase; (xi) ANS: anthocyanidin synthase; (xii) CLL-7: cinnamate–CoA ligase; (xiv) FNS: flavone synthase; (xv) F6H: flavonoid 6-hydroxylase; (xvi) IFS: isoflavone synthase; (xvii) HID: 2-hydroxyisoflavanone dehydratase; (xviii) FNR: flavanone 4-reductase; (xix) F8H: flavonoid 8-hydroxylase; (xx) F3'5'H: flavanone 3′,5′-hydroxylase; (xxi) F3H: flavanone 3-hydroxylase; (xxii) DHK: dihydrokaempferol; (xxiii) DHM: dihydromyricetin; (xxiv) DFR: dihydroflavonol-4-reductase; (xxv) DHQ: dihydroquercetin; (xxvi) FLS: flavonol synthase; (xxvii) OMT: O-methyl transferases; (xxviii) PAL: phenylalanine ammonia lyase; (xxix) UFGT: UDP-glucose flavonoid 3-Oglucosyltransferase; (xxx) LAR: leucoanthocyanidin reductase [21].

2.3.1.1 Phenylpropanoid pathway

The phenylpropanoid route produces flavonoids from phenylalanine, whereas the shikimate pathway produces phenylalanine [22, 23]. The general phenylpropanoid route refers to the first three steps of the phenylpropanoid pathway [24]. The aromatic amino acid phenylalanine is transformed to p-coumaroyl-CoA in this route. The typical phenylpropanoid route begins with the deamination of phenylalanine to trans-cinnamic acid, which is catalyzed by the enzyme phenylalanine ammonia lyase (PAL) [25]. In plants, PAL also has a significant role in controlling the transfer of carbon from primary to secondary metabolism [26]. The second step in the general phenylpropanoid route is catalyzed by the activity of C4H, a cytochrome P450

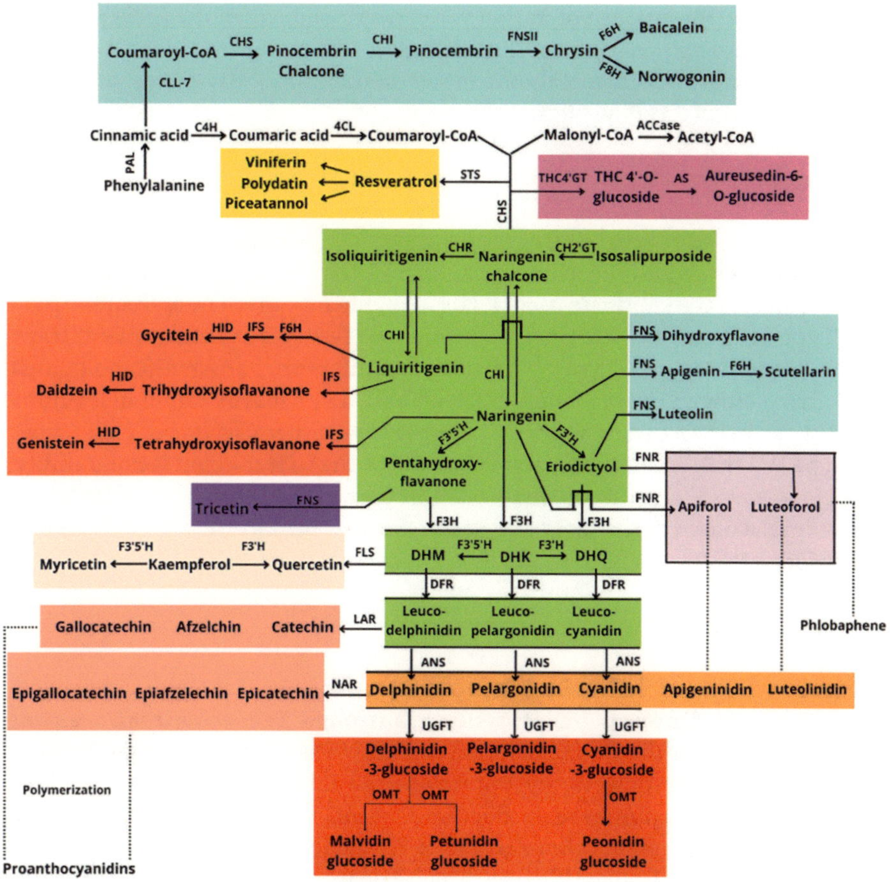

Figure 3.
The branches of flavonoid biosynthetic pathway.

monooxygenase found in plants that hydroxylates trans-cinnamic acid to produce p-coumaric acid [27]. The quantity of lignin, a crucial phenylpropanoid metabolite, in *Populus trichocarpa* and *Arabidopsis thaliana*, is connected to the degree of C4H expression [24, 28]. In the third step, 4-coumarate (4CL) catalyzes the production of p-coumararoyl-CoA by incorporating a coenzyme A (CoA) unit into p-coumaric acid [29].

2.3.1.2 Chalcone biosynthesis

Specific flavonoid synthesis, which starts with chalcone formation, is initiated with the entry of p-coumaroyl-CoA into the flavonoid biosynthesis pathway [30]. One molecule of p-coumaroyl-CoA and three molecules of malonyl-COA are converted into naringenin chalcone (4,2′,4′,6′-tetrahydroxychalcone [THC]) by the action of CHS (produced from acetyl-CoA) [31]. CHS, a polyketide synthase, is the main and first rate-limiting enzyme in the flavonoid biosynthesis pathway [32]. An intermediate of the CHS reaction is subjected to action by the aldo-keto reductase superfamily member chalcone reductase (CHR), which catalyzes its C-6′ dehydroxylation to produce isoliquiritigenin (4,2′,4′-trihydroxychalcone [deoxychalcone]) [33]. In one of the studies, the amount of anthocyanin decreased, when the *Lotus japonicus* CHR1 gene was overexpressed in petunia [21, 34]. Chalcones are recognized as the first important intermediate metabolite in the production of flavonoids, and are also considered as a crucial yellow pigment in plants [35].

2.3.1.3 Flavanones biosynthesis

The intramolecular cyclization of chalcones by CHI, which occurs in the cytoplasm to produce flavanones and the heterocyclic ring C, is a step in the flavonoid pathway [30]. According to the substrate used, CHIs in plants may often be split into two classes. Type I CHIs, which are present throughout the entire vascular plant, transform THC into naringenin. Type II CHIs may manufacture naringenin and liquiritigenin utilizing either THC or isoliquiritigenin and are primarily found in leguminous plants [36]. More than these two forms, there are two other variants of CHI (type III and type IV) that retain the catalytic activity of the CHI fold but lack its ability to cycle chalcones [37]. Additionally, flavanones are a frequent substrate for the downstream flavonoid pathway as well as the flavone, isoflavone, and phlobaphene branches [38, 39].

2.3.1.4 Aurone biosynthesis

Aurones, a family of flavonoids produced from chalcone, are significant yellow pigments in plants. Aurone pigments generate a stronger yellow hue than chalcones and are responsible for the golden coloration of numerous common ornamental plants. Snapdragon, sunflowers, and coreopsis are only a few of the plant species that contain aurones [40, 41]. Aurone production requires THC as a direct substrate [42]. In the cytoplasm of the plant cells, chalcone 4′-O-glucosyltransferase catalyzes the production of THC 4′-O-glucoside from THC. The former is transferred to the vacuole by aureusidin synthase (AS), where it is converted into aureusidin 6-O-glucoside (aurone) [43].

2.3.1.5 Flavone biosynthesis

In all higher plants, flavone production is an essential branch of the flavonoid pathway. Flavone synthase (FNS) converts flavanones into flavones (FNS) [44, 45]. When present in flavanones, FNSI and FNSII encourage the formation of a double bond between C-2 and C-3 positions of the ring C [46]. FNS is a crucial enzyme in the production of flavones. Both naringenin and eriodictyol can be used as substrates by *Morus notabilis* FNSI to produce flavones [47]. Overexpression of *Pohlia nutans* FNSI causes apigenin accumulation in *A. thaliana* [48]. FNSII expression levels in flower buds of *Lonicera japonica* were shown to be congruent with flavone accumulation patterns [46]. Flavanones can be transformed into C-glycosyl flavones as well [21].

2.3.1.6 Isoflavone biosynthesis

Leguminous plants serve as the primary source of isoflavones [49]. Isoflavone synthase (IFS) transports flavanone to the isoflavone route [50] and appears to be able to convert liquiritigenin and naringenin into 2,7,4′-trihydroxyisoflavanone and 2-hydroxy-2,3-dihydrogenistein, respectively [51, 52]. Under the action of hydroxyisoflavanone dehydratase (HID), they are further transformed to the isoflavones genistein and daidzein [53]. Additionally, HID, IFS, and isoflavanone O-methyl transferase can catalyze the conversion of liquiditigenin to 6,7,4′-trihydroxyflavanone, which can then be converted to glycitein (an isoflavone) [54]. IFS and HID catalyze two processes that result in the formation of isoflavone: the formation of a double bond between C-2 and C-3 positions of ring C and the transfer of ring B from C-2 position to C-3 position of ring C [55, 56]. The isoflavone production route begins with IFS, a cytochrome P450 hydroxylase. The accumulation of the isoflavone genistein in invitro tissues was caused by Glycine max IFS overexpression in *Allium cepa* [57]. The use of CRISPR/Cas9 to knock off the expression of the *IFS1* gene resulted in a considerable drop in isoflavones like genistein [44].

2.3.1.7 Flavanol biosynthesis

Flavonols are flavonoid metabolites that have had their ring C-3 hydroxylated [38]. Because their C-3 position is very susceptible to glycosidation, they frequently occur in glycosylated forms in plant cells. Flavonol synthase (FLS) converts the dihydroflavonols like dihydroquercetin (DHQ), dihydrokaempferol (DHK), and dihydromyricetin (DHM) to the flavonols quercetin, kaempferol, and myricetin, respectively [58]. Through the activity of enzymes such as GTs, methyltransferases, and acyltransferase (AT), quercetin, kaempferol, and myricetin are further changed to numerous flavonol derivatives [59]. A C-2 and C-3 double bonds are formed in ring C via the desaturation of dihydroflavonol, which is catalyzed by FLS, a FeII/2-oxoglutarate-dependent dioxygenase. In the flavonol biosynthesis pathway, FLS is considered the key rate-limiting enzyme [21].

2.3.1.8 Anthocyanin and Leucoanthocyanidin Biosynthesis

Major enzyme in flavonoid metabolism in the anthocyanidin and proanthocyanidin pathways is dihydroflavonol-4-reductase (DFR). A hydroxyl group is produced at C-4 position of ring C by the NADPH-dependent reductase known as DFR [60–62]. Dihydroflavonols, DHQ, DHK, and DHM are reduced by DFR to produce leucocyanidin, leucopelargonidin, leucoanthocyanidins, and leucodelphinidin [63]. DFR, for example, transforms DHK to leucopelargonidin in *Vitis vinifera* [64]. The direct synthetic precursor of anthocyanidin and proanthocyanidin. Leucoanthocyanidin, is a crucial intermediary by-product in the flavonoid pathway. The colorless leucopelargonidin, leucocyanidin, and leucodelphinidin are converted into the equivalent anthocyanidins under the catalysis of anthocyanidin synthase (ANS) (the colored pelargonidin, cyanidin, and delphinidin) [65, 66]. An alternative name for ANS is leucoanthocyanidin dioxygenase (LDOX). Similar to FNSI, F3H, and FLS, ANS/LDOX is a FeII/2-oxoglutarate-dependent dioxygenase that stimulates the dehydroxylation of C-4 and formation of a double bond in ring C [67]. In Strawberries, anthocyanin content has been found to get enhanced when ANS is overexpressed [68].

2.3.1.9 Proanthocyanidin biosynthesis

Condensed tannins, also known as proanthocyanidins, are a form of flavonoid made up of leucoanthocyanidins and anthocyanidins [69]. The primary proanthocyanidin units are cis-flavan-3-ols, trans-flavan-3-ols, and flavan-3-ols. Proanthocyanidins are produced when flavan-3-ols are polymerized (or condensed) [70, 71]. To make colored tannins (yellow to brown), polyphenol oxidase (PPO) converts colorless proanthocyanidins into plant vacuoles [72]. The major and rate-limiting enzymes in proanthocyanidin production are leucoanthocyanidin reductase (LAR) and anthocyanidin reductase (ANR). Studies have revealed that overexpression of putative leucoanthocyanidin reductase gene (*PtrLAR3)* significantly elevates proanthocyanidin levels in *Populus tomentosa* [73]. Additionally, in alfalfa (*Medicago sativa*), overexpression of *OvBAN*, an ANR gene obtained from *Onobrychis viviaefolia*, increases the concentration of proanthocyanidin and the activity of the ANR enzyme [74]. However, proanthocyanidin and anthocyanin biosynthesis pathways have a competitive relationship since they utilize the same substrates [75].

2.4 Flavonoid biosynthesis in plants is regulated by transcriptional regulation

In the modification of flavonoid production, transcriptional control is crucial. The major transcriptional regulator in flavonoid biosynthesis is the MBW complex, which consists of WD40, bHLH, and MYB. The MYB domain at the N-terminus of MYB transcription factors (TFs) is needed for DNA binding and interaction with other proteins [76]. According to the amount and location of MYB domain repeats, MYB proteins is categorized into four groups: 3R-MYB, 4R-MYB, R2R3-MYB, and 1R-MYB/MYB-related. Among the four, R2R3-MYB members are mostly engaged in flavonoid metabolism regulation [21].

2.5 Plasticity of flavonoid pathway

Flavonoids have been discovered in epidermal cells such as trichomes, palisade, and spongy mesophyll. Moreover, flavonoids are found intracellularly in numerous cell compartments such as chloroplasts, vacuoles, and the nucleus [77–79]. The shikimate, phenylpropanoid, flavonoid, anthocyanin, and lignin pathways produce plant phenolics. The aromatic amino acids, including phenylalanine, are produced through the shikimate pathway, and the flavonoids are formed by a series of elongation and cyclization stages. Flavonoids get divided into numerous 15-carbon families, including flavanone, flavonol, flavone, flavan-3-ol, anthocyanidin, and isoflavone. It is evident that the level of B-ring hydroxylation is the sole difference between most of the main molecules [80].

2.5.1 Anthocyanin-proanthocyanidin pathway cross-talk

Despite the fact that the anthocyanin and proanthocyanidin routes use identical biochemical intermediates, they are the most and least studied flavonoid processes, respectively. Both branches include the formation of precursors from 4-coumaroyl-CoA and malonyl-CoA. ANS, DFR, and a variety of anthocyanidin-modifying enzymes transform dihydromyricetins, dihydroquercetins, and dihydrokaempferols into anthocyanins. Anthocyanidin rhamnosyltransferases, UDP-glucuronosyl/UDP-glycosyltransferases, methyltransferases, glutathione transferases, and Glycosyltransferases are among the anthocyanidin-modifying enzymes. On the other hand, the family of DFR, LAR, and ANR enzymes convert dihydroflavonols to trans- and cis-epimeric forms of gallocatechins, catechins, and afzelechins in the proanthocyanidin-specific pathway [19, 72, 81, 82].

In a number of plant species, cross-talk between members of the flavonoid pathways' anthocyanin- and proanthocyanidin-specific branches has been seen. Studies have revealed that overexpression of the *ANR* gene in tobacco has suppressed anthocyanin production and induced proanthocyanidin biosynthesis in flower petals. Meanwhile, upregulation of ANR has caused a subset of leaf cells in *Medicago truncatula* plants to produce three times more proanthocyanidin and cut anthocyanin synthesis by half [83, 84].

2.5.2 Lignin-flavonoid pathway cross-talk

Chemical scaffolds of lignin polymers originated and evolved to offer mechanical support to plants, shield them from UV damage and pathogen invasion, as well as increase the hydrophobicity of their vasculature. As a result, these metabolites have played a critical role in the development of land plants and also, in the colonization of various geographical and ecological environments. Similar to flavonoids, this route assisted the manufacture of H and G lignin in early terrestrial plants by enlisting enzymes from primary metabolism [85].

Redirecting metabolic fluxes between the lignin and flavonoid pathways showed molecular and metabolic cross-talks in a variety of plants with down-regulated genes implicated in the phenylpropanoid, lignin, and flavonoid processes. The flow from feruloyl-CoA to G and S units is lessened when the *CCR* gene is silenced in tobacco, tomato, and poplar, which resulted in a decrease in the amount of phenolic chemicals which are particular to lignin [86, 87]. The quantities and composition of several stress-related flavonoid intermediates and derivatives, on the other hand, were significantly increased in these transgenic lines.

3. Effect of gene regulation and modification in flavonoid research and production in crop breeding: recent advances and applications

3.1 Engineering of flavonoid pathway

The flavonoid pathway has been extensively employed in the industry with the goal of accumulating compounds on purpose. Plant species like gerbera, petunia, rose, lisianthus, torenia, and carnation have been genetically modified for the production of novel flower colors. This was achieved by modification of the flavonoid biosynthesis pathway, either via transcriptional down-regulation, inactivation of key anthocyanin pathway enzymes, or by heterologous expression of key enzymes.

There are two significant, possible ways for improving flavonoid biosynthesis. The first is based on the discovery of TFs as a viable alternative to multi-step engineering, while the second is based on the use of inducible promoters to avoid the negative consequences of a constitutive production system. Virus-induced gene silencing has also been proven to be a simple and rapid method of functionalizing TF genes [88].

A few years ago, it was discovered that the pathway to pelargonidin might be opened by transferring a gene encoding DFR from a species where the enzyme does not really exhibit substrate selectivity into a petunia line Lacking F3′5′H activity. In another study, Brick red petunia flowers were produced using the maize gene *A1* and petunia lines with vivid orange blooms obtained from an ornamental plant *Gerbera hybrida* [89]. Now various initiatives are being undertaken to boost anthocyanin concentrations beyond those found naturally. According to a previous short study undertaken, the high-anthocyanin tomatoes have been found to slow tumor development in cancer-prone rats. However, the consequences on human health still require additional research. In this subject, basic proof-of-concept research has been undertaken on a variety of vegetable and fruit species, including apple, grape, tomato, and cauliflower.

3.2 Natural flavonoids variation in horticultural species and horticulture breeding

The molecular basis of various flavonoid production focuses on the activation of genes along with respective pathways by diverse means. The majority of the important structural enzymes, part of the central flavonoid metabolism is encoded by single-copy genes, although some, such as PAL, CHS, F3H, or FLS, are encoded by several genes. The expression of biosynthetic (structural) genes varies significantly between species [19].

Activation of flavonoid and anthocyanin biosynthetic genes in response to light has been reported in most horticultural types. Accumulation of anthocyanin is regarded as one of the most investigated mechanisms in potatoes. This is because colored potato varieties are considered to be a strong source of phytochemicals at levels similar to cranberries, blackberries, blueberries, and grapes. Potato, like other species, has numerous genetic loci that influence anthocyanin production. StAN1, StAN2, StMYBA1, and StMYB113 are important regulators of the phenylpropanoid and anthocyanin pathways. However, bHLH co-factors also play a role, since StAN1 and StAN2 associates with StbHLH1 and StJAF13 in diverse organs, such as the tuber and leaf. A WD40-repeat gene, namely StAN11 has been recently postulated as a regulator of the system via modulating the expression of DFR among other TFs encoding genes, impacting anthocyanin accumulation in potatoes [90]. Apart from potatoes, few other horticulture species have also been exploited due to their antioxidant property via molecular plant breeding techniques [15].

Recently, a variety of red-fleshed and high-flavonoid containing apple genetic resources had been embodied in the complexities of the control of flavonoid production. In one of the studies, red-fleshed apple flavonoid metabolism has been found to get influenced by both hereditary and environmental variables. Numerous flavonoid biosynthesis cascade genes have also been discovered and cloned, so as to identify the flavonoid metabolism that get affected by several environmental factors and genetic variabilities [91].

All these current researches make it evident that flavonoids play a significant role in both food and primary agriculture development and will soon be an intriguing target for molecular plant breeding.

3.3 Flavonoids in tomato breeding

Tomatoes (*Solanum lycopersicum*) are the most abundant dietary source of carotenoids (lycopene), polyphenols, and flavonoids, which are key bioactive compounds favorable to human health. The flavonoids that are mainly produced in tomatoes are predominantly produced mostly in the peels. Naringenin chalcone and rutin (quercetin-rutinoside) are the two primary flavonoids found in tomato fruit so far [92, 93]. To date, three approaches have been made to engineer the flavonoid pathway in tomatoes (*S. lycopersicum*) with the goal to alter its agronomical traits such as its nutritional value, its flower and fruit color as well as its ability to build resistance against insects [94]. They are as follows:

a. Use of structural or regulatory genes to increase endogenous tomato flavonoids; Structural genes are the genes that encode enzymes that directly engage in the synthesis of flavonoids. On the other hand, regulatory genes are the ones that influence the expression of structural genes.

b. Use of RNA interference methods to block particular stages in the flavonoid pathway.

c. Introduction of new flavonoid pathways to produce novel tomato flavonoids [93].

Knowing the fact that there is a lack of flavonoid expression in tomatoes, to date, several attempts have been previously undertaken to generate transgenic tomatoes (**Table 1**).

For example:

i. Heterologous expression of the *FNS-II* gene obtained from *Gerbera* was employed in an attempt to create flavones in tomato. This seemed to have been a good approach to get novel flavone-derived metabolites into tomato fruit [93].

ii. Several components of the MBW complex in tomatoes have recently been discovered and partially described as an anthocyanin biosynthesis regulator. Anthocyanins1 (*SlAN1*) and Anthocyanins2 (*SlAN2*), are the two paralog genes that encode for homologous R2R3-MYB TFs. Both of these paralogs are found in tomatoes on chromosome 10. Thereby, *SIAN1* and *SIAN2* expression in transgenic tomato lines have been found to be responsible for anthocyanin production in multiple organs [92].

iii. To establish hairy root cultures (HRCs), *Agrobacterium rhizogenes* was used to transfer a construct for *PhAN4* expression into the micro tomato genotype MicroTom. HRCs were created to serve as a testbed for whole-plant engineering methodologies that might enhance attributes for space culture [95].

iv. Using TFs to improve a broad spectrum of flavonoids. In addition to the elevated number of flavonoids in tomato fruit caused by *DET1* silencing, the transcription regulator *AtMYB12* expression in tomatoes has been found to trigger flavanol production as well as the caffeoylquinic acid biosynthetic pathway [96].

Crop Variety/Name	Gene responsible	Name of Flavonoid Induced	Reference
Tomato	*FNS-II* gene of *Gerbera*	N/A	[93]
Tomato	*SlANT1* and *SlAN2* (coding for R2R3-MYB)	Anthocyanin	[92]
Tomato	*CHI* gene from red onion by transgenesis	Flavanol and Anthocyanin	[94]
Tomato	*Del/Ros1* gene from snapdragon by transgenesis	Flavanol and Anthocyanin	[94]

Table 1.
Genes responsible for different flavonoid production in tomato.

v. The *CHI* gene from red onion and the *Del/Ros1* gene from snapdragon has been employed to boost the flavanol and anthocyanin content of tomatoes, respectively, via transgenic techniques [94].

vi. In one of the studies, simultaneous overexpression of the two maize TFs C1 and Lc have resulted in a 60-fold upregulation in kaempferol glycosides in tomato flesh tissue [96].

In one of the studies, it was concluded that despite having intense debate over the advantages, disadvantages, and risks of genetically modified food, around 96% of the customers showed interest in purchasing high flavonoid containing tomatoes. It is considered that this changing mindset of people will prove to be crucial for the development of transgenic vegetables in the future [94]. Moreover, various other studies are in process to make transgenic tomato breeding more productive and nutritional in the coming future.

3.4 Flavonoids in rice breeding

Rice is staple food in many Asian countries. Even though white rice is the most popular, Asian cuisine often includes colored rice. Several pieces of evidence reveal that pigmented rice has important biological properties, including antioxidants, anti-allergic, and neuro-protective properties. The rich flavonoid and nutritional content of colored rice warrants enhancement flavonoid content in rice by implementation of different breeding strategies (**Table 2**).

The functional activities of TFs influence the color of rice grains. In black rice, the *Kala3* gene, which codes for R2R3-Myb, and the *Kala4* gene, which codes for basic helix–loop–helix (bHLH), activate the flavonoid biosynthesis genes *ANS CHS*, and *DFR* resulting in anthocyanin pigment buildup in the grain. In red rice, the *Rc* gene expressing bHLH activates *CHS*, *DFR*, and *LAR*, resulting in the buildup of proanthocyanidin pigment in the grain. The promoter of *Kala4* in white rice differs from that in pigmented rice, and loss of 14 base pairs inside the *Rc* open reading frame, resulted in lack of color in the grain. In addition, the gene *CYP75B3* is strongly expressed in pigmented rice grains, along with other flavonoid pathway genes. This explains why leucocyanidin-derived anthocyanin and proanthocyanidin pigments are abundant in colored rice grains [97].

Recently, in order to create flavone, isoflavone, and flavonol in rice grain, the flavonol (*AtF3H/AtFLS*), isoflavone (*GmIFS*), and flavone (*PoFNSI/GmFNSII*) biosynthetic enzyme genes, as well as *OsPAL* and *OsCHS*, were expressed in a

Crop Variety/ Name	Gene responsible	Codes for	Name of Flavonoid Induced	Reference
Black Rice	*Kala3*	R2R3-Myb	N/A	[97]
Black Rice	*Kala4*	bHLH	Anthocyanin	[97]
Red Rice	*Rc*	bHLH	Proanthocyanidin	[97]
Rice	*OsCOP1* gene introduced via CRISPR-Cas9	N/A	N/A	[98]

Table 2.
Genes responsible for different flavonoid production in rice.

seed-specific way. These biosynthetic genes were expressed in seed using the *GluB-1* promoter and the 18-kDa oleosin promoter [99]. In another study, the *OsCOP1* gene, an ortholog of *Arabidopsis thaliana* constitutive photomorphogenic 1 (COP1) was introduced in rice using CRISPR-Cas9. This not only turned the pericarp of the rice variety yellowish but also caused embryonic death. Moreover, this also reduced the size of the transgenic seeds [98]. According to a study, a total of 82 flavonoids have been chemically identified in transgenic rice seeds. Moreover, exogenous enzymes produced flavonoids in rice seeds that were later altered by endogenous enzymes and transported, causing persistent accumulation in PB-I and/or PB-II. Based on these results, the heterologous and ectopic expression of biosynthetic enzymes in rice seeds not only serves as a productive platform for the production of flavonoids but can also be used to broaden the structural diversity of flavonoids and hence open up a new, untapped source of bioactive substances [100].

Increasing the flavonoid biosynthesis and its accumulation in rice have been found to contribute to the enhanced heat tolerance under stress, as well as plays a regulatory role in the activation of the antioxidant enzyme system [101].

3.5 Flavonoid research in maize

Obtaining security of grain supply in the twenty-first century with limited arable land is a big challenge because of the constantly changing environment and increasing global population [102–104]. Maize plays a very important role in global grain production. Drought is a significant factor restricting plant development and productivity. Drought stress affects growth and development of plants, which is directly related to yield. Under drought stress, *doi57* gene is observed to play an important role in maintaining the plant to grow and survive in order to give good yield. *doi57* gene is one of the key genes involved in biosynthesis of flavonoid. With less soil water content (SWC), doi57 guard cells can accumulate more flavonols and less hydrogen peroxide (H_2O_2). Furthermore, under drought conditions, *doi57* seedling extracts had a stronger potential to scavenge oxygen free radicals than B73 maize genome. Moreover, in terms of transpiration rates, photosynthetic rates, water consumption efficiency, and stomatal conductance, *doi57* seedlings outperformed B73, resulting in high biomass and enhanced root/shoot ratios in *doi57* mutant plants [105].

3.6 Flavonoid profile in millets breeding

Millet polyphenols protect the neurological system by lowering oxidative stress, and vitexin is a crucial component of millet polyphenols. Vitexin, a flavonoid derived from millet, is present in many foods such as millet, mung bean, and others. It is also known chemically as apigenin-8-C-glucoside. According to research, vitexin contains potent free radical scavenging and antioxidant enzyme protection properties that may protect cells from oxidative damage [106].

3.7 Flavonoids in olive breeding

With almost 1200 olive varieties listed, the olive has a great genetic diversity. High heterozygosity, prolonged juvenile phase, and a paucity of information on trait heritability have all been major limiting factors in olive breeding. Attempts to obtain new varieties have concentrated on improving olive response to varied growing situations through systematic breeding. New olive breeding methods uses two varieties "Picual" and "Arbequina" as controls along with other varieties. The phenolic compound metabolism in the olive tree is quite complex, and is controlled by environmental and genetic factors that regulate the final phenolic composition of olive fruits. The unique breeding selection UCI2–68 demonstrated an optimal phenolic profile, resulting in good agronomic performance [107].

3.8 Flavonoids in soybean breeding

In adverse conditions, soybean (*Glycine max*) productivity drops significantly. Soybean breeding might benefit from discovering regulatory components that impart stress tolerance. *HSFB2b*, a class B heat shock factor, enhances salt tolerance by activating one subset of flavonoid biosynthesis-related genes and blocking the repressor *GmNAC2* to release another collection of flavonoid biosynthesis-related genes. Silencing *GmFNSII*, the principal flavone-producing gene, reduces flavone levels and increases salt sensitivity in hairy soybean roots [108]. Leaf-chewing insects are severe pests of soybeans, lowering seed quality and limiting output (*G. max*). The CRISPR/Cas9 expression vector was introduced into the soybean cultivar via agrobacterium-mediated transformation, resulting in *Glyma.07 g110300*-gene mutants. A 33-bp deletion and a single-bp insertion in the *GmUGT* coding domain increased resistance to Cotton bollworm (*Helicoverpa armigera*) and Tobacco cutworm (*Spodoptera litura*). Furthermore, *GmUGT* overexpression made soybean species susceptible to *H. armigera* and *S. litura* [109].

3.9 Molecular breeding of peanut with high flavonoid content

Peanuts include a variety of bioactive substances in addition to helpful fatty acids and minerals. Flavonoids found in peanuts include flavonol, dihydroquercetin, C-glycoside flavone, dihydroflavonol, flavonone, and 5,7-dimethoxyisoflavone. BARI2011 is the most drought-tolerant of the peanut cultivars. According to one study, BARI2011 retained more water. MYB123 encodes an R2R3 MYB domain-containing TF that has been demonstrated to upregulate flavonoid production and is a critical determinant in proanthocyanidin accumulation. Correlational investigation of TF expression with flavonoid biosynthesis and accumulation of phenolics, flavanols, and anthocyanins in peanuts revealed that TFs coregulated flavonoid production under water stress [110].

3.10 Flavonoid content in mustard

The seed coat color of Brassica crops is an essential horticultural feature. Seeds with a yellow seed coat have higher oil quality, more protein, and less fiber. As a result, the yellow seed coat color is seen as a good characteristic in *Brassica juncea, Brassica rapa*, and *Brassica napus* hybrids. The majority of seed coat color is produced by the accumulation of proanthocyanidins, the ultimate product of the flavonoid biosynthetic pathway, which is mostly deposited in the innermost cell layer of testa (chalaza, micropyle, and endothelium) [111].

3.11 Flavonoid content in lettuce breeding

Lettuce (*Lactuca sativa*) is one of the world's most important vegetables. The GWAS identified 5311 expression quantitative trait loci (eQTL) that influence the expression of 4105 genes, including nine eQTLs that regulate flavonoid biosynthetic genes. GWAS has found six candidate loci for anthocyanin variation in lettuce leaves [112].

UV-A supplementation increased flavonoids, anthocyanin, and polyphenol levels and the 2,2-diphenyl-1-picryl-hydrazyl-hydrate (DPPH) free radical scavenging rate. UV-A can modify plant phenolic contents and flavonoid metabolism by increasing the expression of associated genes such as *CHS* and *MYB* in the flavonoid pathway and *PAL* in the propane metabolic pathway. Under additional UV-A and FR light, anthocyanin levels in lettuce seedlings were 11% higher and 40% lower, respectively [113] (**Table 3**).

Transformation of *L. sativa L.* with *rol C* gene inducing an increase in total flavonoid contents [114].

The gene expression of rol ABC genes was used to boost secondary metabolites in *L. sativa L.* (cv. Grand Rapids), particularly antioxidants such as phenolics and flavonoids. *A. tumefaciens* GV3101 and the *rol ABC* genes were used to transform *Lactuca sativa L.* (cv. Grand Rapids). The transformation increased the secondary metabolites in lettuce and also induced free radical inhibitor effect and lipid peroxidation scavenging properties [115].

Thereby, the three key categories that is used to group the goals of current lettuce breeding programmes include: (1) Improvement in horticultural traits such as quality and resistance to early bolting, (2) resilience to diseases and pests, and (3) to attain higher yield and uniformity [116].

3.12 Flavonoid content in buckwheat

Buckwheat (*Fagopyrum esculentum*) is an annual crop that is planted all over the world. Buckwheat seeds, leaves, and stems are high in flavonoids such as rutin and proanthocyanidins (PAs). The discovery of the ANR and LAR buckwheat genes could result in the development of buckwheat cultivars with different PA levels. In one investigation, one gene sequence (*AT1*, Fes sc0002933.1.g000003.aua.1) encoding ANR was found, along with three gene sequences encoding LARs (*LT1*, Fes sc0001063.1.g000007.aua.1; *LT2*, Fes sc0016501.1.g000002.aua.1; and *LT3*, Fes sc0010963.1.g000003.aua.1) [117].

The flavonoid contents in buckwheat sprouts were commonly in the following order: rutin > quercetin > isovitexin > vitexin > isoorientin > kaemferol [118]. Tartary buckwheat grain contains orientin, vitexin, rutin, and quercetin and is used to make drinks and biscuits. Because of its high content of flavonoids and other phenolic compounds, tartary buckwheat is resistant to pests, plant diseases, and UV-B radiation damage. As a result, Tartary buckwheat can be cultivated organically and without synthetic fertilizers or chemical treatments [119].

Crop Variety/ Name	Gene used	Name of Flavonoid Induced	Method Used	Reference
Lettuce	*rol C*	N/A	Agrobacterium-mediated transformation	[114]
Lettuce	*rol ABC*	N/A	Agrobacterium-mediated transformation	[115]

Table 3.
Genes responsible for different flavonoid production in Lactuca sativa L.

3.13 Distinctive flavonoid profiles in legume

Legumes have sparked a lot of attention because of their health benefits and their polyphenolic compounds. The xanthine oxidase XO inhibitory activity appears to be reduced when a glycoside or a methyl group is substituted for the hydroxyl groups at C7 and C3 of the basic flavonoid structure [120].

Recently, isolation of genes encoding the critical enzymes of various phenylpropanoid branch pathways has opened the door for engineering crucial agricultural plants like alfalfa for:

a. improving the forage digestibility through lignin composition and content,

b. enhancing the disease resistibility via introducing novel phytoalexins, or by changing transcriptional regulator expressions,

c. improving nodulation efficiency by overproduction of flavonoid nod gene inducers [121].

The recognition of flavonoids in legume root exudates by the bacterium, originates in the rhizosphere and activates a particular set of genes involved in bacterial nodulation (*Nod*). These nodulations produce and secretes a very specific signal known as lipochito-oligosaccharides or Nod factors. The formation of rhizobia-infected root nodules is the result of a variety of host responses brought on by the perception of Nod factors by the plant LysM receptors, including bacterial invasion of the root hairs, root hair curling, cortical cell divisions and induced expression of the host symbiotic genes [122].

4. Other functions and applications of flavonoids

4.1 Effects and influence of flavanol rich cocoa on cognitive performance

The potential of flavonoids to alter signaling pathways enhancing neuronal function and brain connection appears to be one of the processes underlying the health benefits of frequent cocoa consumption [123].

4.2 Role for polyphenols in the prevention of degenerative diseases

Flavonoids protect the brain cells in a multitude of ways, such as by strengthening the functional neurons or by promoting neuronal regeneration. After 6-hydroxydopamine lesioning, it has been found that the citrus flavanone tangeretin preserves nigro-striatal integrity and functioning in the context of Parkinson's disease [124].

4.3 Flavonoids in decorticated sorghum grains exert antioxidant and antidiabetic activities

Flavonoids are said to be helpful in preventing metabolic disorders including type 2 diabetes, obesity, hypertension, and several malignancies. They are primarily abundant in the bran portion (pericarp, testa, and aleurone tissues) of the plants. According to the findings, decorticated sorghum grains contain significant amounts of flavonoids, which may inhibit glucose hydrolyzing enzymes and lessen the symptoms of diabetes and its companion disorders [125].

4.4 Chrysoeriol7: a natural chemical and repellent, against brown planthopper in rice

Brown Planthopper (BPH), a rice pest, causes severe damage such as viral infections, leaf blights, nutrition loss, and tissue death, all of which have a direct impact on rice productivity. Chrysoeriol7, a secondary metabolite, has been reported to be effective against BPH by establishing resistance due to the presence of an aromatic group from the flavonoid family that emanates a distinct scent that repels BPH [126].

4.5 Quinoa-derived polyphenols regulate glucose and lipid metabolism: protecting against chronic human illnesses

Quinoa is known as the "golden grain" because of its excellent protein profile, high polyphenol and vitamin content, and health advantages such as anti-diabetic, antioxidant, and anti-obesogenic properties. Furthermore, quinoa leaves are high in phenolic compounds, which can help reduce the risk of cardiovascular disease, neurological illnesses, and diabetes. Quinoa extracts contains significant quantities of sinapinic, ferulic, and gallic acids, isorhamnetin, kaempferol, and rutin. These chemicals are associated directly to a reduction in prostate cancer cell growth and motility [127, 128].

4.6 *B. juncea* (L.) Czern. leaves show high flavonoid content: reducing rheumatoid arthritis caused by adjuvants

The leaves of *B. juncea* contain certain medicinal components that have been found to reduce the synovial inflammation as well as treat the damage caused by rheumatoid arthritis [129].

4.7 Nutritional value of Tartary buckwheat for humans and activity evaluation of its major flavonoids

Tartary buckwheat, which is produced mostly in northern India, Nepal, Bhutan, China, and central Europe, has been found to be more cold resistant and drought tolerant than regular ones. The phenolic compounds, resistant starch, and protein contained in grains, as well as interactions between these constituents, are largely responsible for reduction of the risk of a number of chronic illnesses such as cardiovascular disease, obesity, hypertension, and gallstone formation. Tartary buckwheat is resistant to pests, plant diseases, and UV-B radiation because it contains flavonoids such as rutin, vitexin, orientin, and quercetin, as well as other phenolic compounds [119].

4.8 Using purple tomato anthocyanins as new antioxidants to improve human health

It is generally known that tomatoes help to lessen the risk of developing cancer since they contain carotenoids and polyphenols. According to several recent research, anthocyanins can have a variety of impacts on the health of the eye.

Recently, in order to examine the potential impacts of genetically modified "Indigo" tomatoes on the host gut microbiota, inflammatory reactions and the signs of inflammatory bowel diseases (IBDs) were tested in an unexpected ulcerative colitis mouse model. In addition, consumption of anthocyanins, in particular, has shown a favorable correlation with a decrease in cardiovascular risk, as well enhanced the vascular health and prevented the formation of atherosclerotic plaque [130].

5. Effect of gene regulation and modification in Flavonoid production

5.1 A citrus cytochrome P450 gene, *CsCYT75B1* helping to induce drought tolerance by antioxidant Flavonoid accumulation

Cytochrome P450 gene present in *Citrus sinensis* (*CsCYT75B1*) is linked with flavonoid metabolism and reported to be notably induced after drought stress. *CsCYT75B1* gene when overexpressed in *A. thaliana* significantly increased total flavonoid content. It also enhanced antioxidant activity in the transgenic Arabidopsis plant. Diverse genes responsible for flavonoids biosynthesis showed increased induction (2–12 folds) as a result of *CsCYT75B1* gene overexpression in these transgenic Arabidopsis lines. After induction of draught stress, these plants showed enhanced drought tolerance along with antioxidant flavonoids accumulation, lower level of ROS and superoxide radicals when compared to wild type plants. These transgenic lines also exhibited significantly lower levels of electrolytic leakage than wild types [131].

5.2 Selected mutagenesis of *GmUGT* augmented soybean resistance against leaf-chewing insects via flavonoids Biosynthesis

Seed quality and yield in Soybeans (*G. max*) is negatively affected by Leaf-chewing insects. Minimization of insecticide use and loss reduction can be achieved by breeding Leaf-chewing insects-resistant soybean varieties. Marker genes for QTL-M, *Glyma.07g110300* (LOC100775351) encoding UDP-glycosyltransferase (UGT) is the major deciding factor for resistance against leaf-chewing insects in soybean. It manifests loss of function in insect-resistant germplasms of soybean. Zhang Y et al. reported a study, where they have introduced CRISPR/Cas9 expression vector into the soybean cultivar Tianlong No. 1 using Agrobacterium-mediated transformation to generate *Glyma.07g110300*-gene mutants. A 33-bp deletion and a single-bp insertion in the *GmUGT* coding region obtained from this experiment resulted in enhanced resistance to *Helicoverpa armigera* and *S. litura*. Soybean varieties further sensitive to *H. armigera* and *S. litura* was generated by upregulation of *GmUGT*. This particular coding sequence is also involved in providing resistance to leaf-chewing insects via alteration of flavonoid content and gene expression pattern related to flavonoid biosynthesis and defense [109].

5.3 Transferability and polymorphism of simple sequence repeats (SSRs) In the Flavonoid pathway genes of strawberry (*Fragaria*) and Rasberry (*Rubus Sp.*)

Fragaria and *Rubus* are two extremely popular crops, whose breeding programs are increasingly reliant on the use of functional DNA markers. Improvement in the nutritional quality and disease resistance abilities are essential challenges in breeding programs of these crops. There is a total of 118 microsatellite (simple sequence repeat-SSR) loci in the nucleotide sequences of genes that are involved in flavonoid production and pathogenicity and a count of 24 SSR markers that represent some of these structural and regulatory genes have yet been discovered. Using these markers, 48 specimens of *Fragaria* and *Rubus*, comprising unusual cultivars and wild species were examined in one of the studies to determine their overall genetic diversities. It is believed that the enhancement of anthocyanin-related phenotypes in strawberry and raspberry breeding programs may get benefited from the use of SSR markers collection as a molecular tool [132].

5.4 *B. napus L.'s* efficient oil production is controlled by targeted mutation of *BnTT8* homologs

B. napus is an important oil crop, but despite its importance, no naturally occurring or artificially produced yellow seed germplasms have been discovered yet. Recently, in one of the studies, CRISPR/Cas9 system has been used to produce yellow mutant rapeseeds (*Brassica napus*). The targeted alterations of the *BnTT8* gene were persistently passed down through generations, and a variety of homozygous mutants with damaged alleles of the target genes were acquired for phenotyping. *BnA09.TT8* and *BnC09.TT8b* are the two targeted mutants of *BnTT8* which were able to restore the yellow-seeded phenotype. These mutants generated seeds with increased protein and oil content, improved fatty acid (FA) composition, and no significant abnormalities in yield-related parameters [133].

6. Mediterranean crop modification techniques: from the laboratory to the field

6.1 Marker-assisted breeding (MAB)

This method/strategy selects plants and animals for breeding programs early in their development by exploiting DNA markers linked with desirable features. Thus, it significantly shortens the time required in a breeding cycle, to locate/identify variations or breeds that display the desired trait. Two separate studies in tomatoes recorded before, one using a mutant inbred line and the other using an interspecific *Solanum Chmielewski* population, discovered that the colorless-peely mutant on chromosome 1 is controlled by a *SlMYB12*-regulated transcriptional network that controls the accumulation of yellow-colored flavonoid (naringenin chalcone) in the fruit epidermis [134, 135].

6.2 Crop plants that have undergone genetic modification

Overexpression and competition with the target route, overriding rate-limiting steps, preventing the catabolism pathway of the desired product, and blocking other pathways are all part of the process of maximizing the synthesis of specialized target molecules [136]. For example, the principal flavonoid metabolic route has been studied to optimize these critical molecules in *Solanum lycopersicum*. Flavonoids (such as naringenin, chalcone, and rutin) are predominantly found in tomato peel, with just trace levels found in tomato flesh [137]. Ectopic expression of a single structural gene (*CHI*) or many structural genes (*CHS*, *CHI*, *F3H*, and *FLS*) increased the number of flavonols (quercetin- and kaempferol- glycosides) in tomato peel and flesh. The co-expression of onion *CHI* in the purple tomato *Delila* bHLH and *Rosea1* R2R3-MYB recently transformed flavonoid to flavanol and boosted anthocyanin concentration [138].

6.3 Modern breeding methods (MBTs)

The rapid growth and usage of genome editing tools has opened up new ways for introducing change and influencing gene expression at multiple levels, including transcription, mRNA processing, and mRNA translation. CRISPR/Cas9 technology has primarily been used to study plant biosynthetic potential by blocking competing biosynthetic pathways and changing metabolite flux towards target chemical synthesis. One such example, the Salvia miltiorrhiza rosmarinic acid synthase (*SmRAS*) gene was edited using the CRISPR/Cas9 technology [139]. This mutation led to a decrease in phenolic acid content, such asrosmarinic acid, and an increase in its precursor, 3,4-dihydroxyphenyl lactic acid, especially in the homozygous variety. Another study in *S. miltiorrhiza* employed CRISPR/Cas9 to knock out the *SmCPS1* gene, which codes for a diterpene synthase involved in tanshinone synthesis. This was done to investigate the feasibility of encouraging the accumulation of the substrate for taxol synthesis as tanshinones and taxol sharing the same precursor (Geranylgeranyl Pyrophosphate) [140].

6.4 Regulation of specific metabolism and transcription factor modulation

The functional characterization of TFs involved in the control of anthocyanin metabolism is an outstanding example of how transcriptional regulatory research can be carried out to fine-tune specialized metabolic pathways. The WD-repeat/bHLH/MYB complex regulates anthocyanin accumulation in plants by positively regulating the gene expression of DFR, anthocyanin synthase, and glucosyltransferase. This molecular pathway, conserved across many species via orthologous TFs, which is essential in the color determination of flowers and fruits such as those of apples, grapes, and oranges [141].

6.5 A target for molecular biology-based breeding and other biotechnological approaches: metabolism of plant glandular Trichomes

Trichomes are specialized biosynthetic, storage structures composed of epidermal extensions found on the surface of aerial plant parts. They can exist in both non-glandular and glandular trichrome forms and are widespread throughout the plant kingdom. Glandular trichomes can produce, store, and release exudates containing a wide range of chemo-diverse compounds such as essential oils, oleoresins, phenols, glycerids, and extremely complex terpenes [142]. Efficient isolation techniques such as laser microdissection pressure catapulting (LMPC) has aided with the separation and enrichment of specific cell types, such as multicellular glandular trichomes allowing chemical, transcriptional, and biosynthetic studies to focus solely on specialized glandular metabolites [143].

7. Future prospect

Flavonoid-rich diets have been shown to delay the onset of dementia-related diseases and prevent age-related cognitive decline. Alterations in cerebral blood flow caused by flavonoids, progenitor cells, quantitative changes in brain stem cells and gray matter density, as well as electrophysiological anomalies, can all be examined utilizing imaging and spectroscopic tools such as NMR and MRI. All of these initiatives will result in mechanism-based linkages between flavonoid medication and brain activities, as well as therapeutic dosage data. Flavonoids have been shown to concentrate in the brain and activate Akt-CREB and ERK-CREB mediated memory, making them interesting therapeutic candidates for memory enhancement. Flavonoid-loaded nanoparticles, liposomes, or other nanocarriers can pave the way for flavonoids in the future by increasing the half-life of flavonoids in organisms, drug delivery strategies boost their effects. Moreover, since flavonoids are naturally occurring dietary components, therapeutically effective amounts of flavonoids can have varying lethal effects on cancer cell lines as well as tissues due to cancer heterogeneity. Ototoxicity is a side effect of chemotherapeutic medicines that patients experience during chemotherapy. According to one study, epigallocatechin-3-gallate protects patients against ototoxicity, which can also pave the door for flavonoids to be utilized as additive treatments to reduce the side effects of chemotherapy drugs. It is predicted that addressing difficulties like as bioavailability and metabolism, building physiologically acceptable in vitro models, determining the effects of processing, standard measuring methodologies, and adequate clinical biomarkers will surely influence the future of flavonoid research [144–146].

8. Conclusion

The use of phytochemicals, particularly flavonoids, in disease prevention and treatment is well documented. Each flavonoid discovered in nature has distinct chemical, physical, and physiological features. The structure-function link of flavonoids is the pinnacle of major biological functions. Incorporation of flavonoid in various horticulture species and the practice of molecular plant breeding will bring in a great revolution in the coming future. This will elevate numerous therapeutic potentials of several plant species and would thereby provide aid in the advancement of the food, pharmaceutical, floricultural, and chemical industries. Additional accomplishments will bring in other newer insights and will almost probably usher in a new era of flavonoid-based pharmacological agents for the treatment of numerous infectious and degenerative disorders.

Author details

Shuchi Nagar[1†], Saurabh Dey[1†], Aishik Das[2] and Soumya Basu[2*]

1 Bioinformatics Research Laboratory, Dr. D Y Patil Biotechnology and Bioinformatics Institute, Dr. D Y Patil Vidyapeeth, Tathawade, Pune, India

2 Cancer and Translational Research Centre, Dr. D Y Patil Biotechnology and Bioinformatics Institute, Dr. D Y Patil Vidyapeeth, Tathawade, Pune, India

*Address all correspondence to: soumya.bs@gmail.com

† Equal Contribution.

References

[1] Panche AN, Diwan AD, Chandra SR. Flavonoids: An overview. Journal of nutritional. Science. 2016;**5**

[2] Heim KE, Tagliaferro AR, Bobilya DJ. Flavonoid antioxidants: Chemistry, metabolism and structure-activity relationships. The Journal of Nutritional Biochemistry. 2002;**13**(10):572-584

[3] Pietta PG. Flavonoids as Antioxidants. Journal of Natural Products. 2000;**63**(7):1035-1042

[4] Havsteen BH. The biochemistry and medical significance of the flavonoids. Pharmacology & Therapeutics. 2002;**96**(2-3):67-202

[5] Hatti KS, Diwakar L, Rao GV, Kush A, Reddy GC. Abyssinones and related flavonoids as potential steroidogenesis modulators. Bioinformation. 2009;**3**(9):399-402

[6] Ullah A, Munir S, Badshah SL, Khan N, Ghani L, Poulson BG, et al. Important flavonoids and their role as a therapeutic agent. Molecules. 2020;**25**(22):5243

[7] Nagao A, Seki M, Kobayashi H. Inhibition of xanthine oxidase by flavonoids. Bioscience, Biotechnology, and Biochemistry. 1999;**63**(10):1787-1790

[8] Hanasaki Y, Ogawa S, Fukui S. The correlation between active oxygens scavenging and antioxidative effects of flavonoids. Free Radical Biology and Medicine. 1994;**16**(6):845-850

[9] Tsukasa I. Contribution to flower colors of flavonoids includinganthocyanins: a review. Natural Product Communications. 2015;**10**(3):529-544

[10] Photoprotection of natural flavonoids. Journal of applied pharmaceutical Science. 2013

[11] Dudek B, Warskulat AC, Schneider B. The occurrence of flavonoids and related compounds in flower sections of Papaver nudicaule. Plants. 2016;**5**(2):28

[12] Kumar S, Pandey AK. Chemistry and biological activities of flavonoids: An overview. The Scientific World Journal. 2013;**2013**:1-16

[13] Mathesius U. Flavonoid functions in plants and their interactions with other organisms. Plants. 2018;**7**(2):30

[14] Laoué J, Fernandez C, Ormeño E. Plant flavonoids in Mediterranean species: A focus on Flavonols as protective metabolites under climate stress. Plants. 2022;**11**(2):172

[15] D'Amelia V, Aversano R, Chiaiese P, Carputo D. The antioxidant properties of plant flavonoids: Their exploitation by molecular plant breeding. Phytochemistry Reviews. 2018;**17**(3):611-625

[16] Yonekura-Sakakibara K, Higashi Y, Nakabayashi R. The origin and evolution of plant Flavonoid metabolism. Frontiers. Plant Science. 2019;**10**

[17] Gensheimer M. Chalcone isomerase family and fold: No longer unique to plants. Protein Science. 2004;**13**(2):540-544

[18] Markham KR. Distribution of flavonoids in the lower plants and its evolutionary significance. 1988:427-468

[19] Flavonoid W-SB, Biosynthesis. A colorful model for genetics,

biochemistry, cell biology, and biotechnology. Plant Physiology. 2001;**126**(2):485-493

[20] Ferreyra MLF, Rius SP, Casati P. Flavonoids: Biosynthesis, biological functions, and biotechnological applications. Frontiers. Plant Science. 2012;**3**

[21] Liu W, Feng Y, Yu S, Fan Z, Li X, Li J, et al. The Flavonoid Biosynthesis network in plants. International Journal of Molecular Sciences. 2021;**22**(23):12824

[22] Wang ZL, Wang S, Kuang Y, Hu ZM, Qiao X, Ye M. A comprehensive review on phytochemistry, pharmacology, and flavonoid biosynthesis of iScutellaria baicalensis/i. Pharmaceutical Biology. 2018;**56**(1):465-484

[23] Rehan M. Biosynthesis of Diverse Class Flavonoids Ivia/i Shikimate and Phenylpropanoid Pathway. 2021

[24] Dong NQ, Lin HX. Contribution of phenylpropanoid metabolism to plant development and plant–environment interactions. Journal of Integrative Plant Biology. 2021;**63**(1):180-209

[25] Williams JS, Thomas M, Clarke DJ. The gene stlA encodes a phenylalanine ammonia-lyase that is involved in the production of a stilbene antibiotic in Photorhabdus luminescens TT01. Microbiology. 2005;**151**(8):2543-2550

[26] Barros J, Dixon RA. Plant phenylalanine/tyrosine ammonia-lyases. Trends in Plant Science. 2020;**25**(1):66-79

[27] Wohl J, Petersen M. Functional expression and characterization of cinnamic acid 4-hydroxylase from the hornwort Anthoceros agrestis in Physcomitrella patens. Plant Cell Reports. 2020;**39**(5):597-607

[28] Fraser CM, Chapple C. The Phenylpropanoid pathway in Arabidopsis. The Arabidopsis Book. 2011;**9**:e0152

[29] Li Y, Kim JI, Pysh L, Chapple C. Four isoforms of Arabidopsis thaliana 4-coumarate: CoA ligase (4CL) have overlapping yet distinct roles in phenylpropanoid metabolism. Plant Physiology. 2015;**2015**:00838

[30] Nabavi SM, Šamec D, Tomczyk M, Milella L, Russo D, Habtemariam S, et al. Flavonoid biosynthetic pathways in plants: Versatile targets for metabolic engineering. Biotechnology Advances. 2020;**38**:107316

[31] Wang J, Li G, Li C, Zhang C, Cui L, Ai G, et al. NF-Y plays essential roles in flavonoid biosynthesis by modulating histone modifications in tomato. New Phytologist. 2020;**229**(6):3237-3252

[32] Deng X, Bashandy H, Ainasoja M, Kontturi J, Pietiäinen M, Laitinen RA, et al. Functional diversification of duplicated chalcone synthase genes in anthocyanin biosynthesis of Gerbera hybrida. New Phytologist. 2013;**201**(4):1469-1483

[33] Bomati EK, Austin MB, Bowman ME, Dixon RA, Noel JP. Structural elucidation of Chalcone reductase and implications for Deoxychalcone Biosynthesis. Journal of Biological Chemistry. 2005;**280**(34):30496-30503

[34] Shimada N, Nakatsuka T, Nishihara M, Yamamura S, Si A, Aoki T. Isolation and characterization of a cDNA encoding polyketide reductase in Lotus japonicus. Plant Biotechnology. 2006;**23**(5):509-513

[35] Ohno S, Yokota M, Yamada H, Tatsuzawa F, Doi M. Identification of Chalcones and their contribution

to yellow coloration in dahlia (Dahlia variabilis) ray florets. The Horticulture Journal. 2021;**90**(4):450-459

[36] Yc Y, Zhang X, Gao Zq HT, Liu Y. The research Progress of Chalcone isomerase (CHI) in plants. Molecular Biotechnology. 2018;**61**(1):32-52

[37] Chao N, Rf W, Hou C, Yu T, Miao K, Fy C, et al. Functional characterization of two chalcone isomerase (CHI) revealing their responsibility for anthocyanins accumulation in mulberry. Plant Physiology and Biochemistry. 2021;**161**:65-73

[38] Lin S et al. R2R3-MYB transcription factors, StmiR858 and sucrose mediate potato flavonol biosynthesis. Horticulture Research. 2021;**8**(1)

[39] Casas MI, Falcone-Ferreyra ML, Jiang N, Mejía-Guerra MK, Rodríguez E, Wilson T, et al. Identification and characterization of maize isalmon silks/i genes involved in insecticidal Maysin Biosynthesis. The Plant Cell. 2016;**28**(6):1297-1309

[40] Alsayari A, Muhsinah AB, Hassan MZ, Ahsan MJ, Alshehri JA, Begum N. Aurone: A biologically attractive scaffold as anticancer agent. European Journal of Medicinal Chemistry. 2019;**166**:417-431

[41] Nakayama T, Yonekura-Sakakibara K, Sato T, Kikuchi S, Fukui Y, Fukuchi-Mizutani M, et al. Aureusidin synthase: A polyphenol oxidase homolog responsible for flower coloration. Science. 2000;**290**(5494):1163-1166

[42] Ono E, Fukuchi-Mizutani M, Nakamura N, Fukui Y, Yonekura-Sakakibara K, Yamaguchi M, et al. Yellow flowers generated by expression of the aurone biosynthetic pathway. Proceedings of the National Academy of Sciences. 2006;**103**(29):11075-11080

[43] Ono E, Hatayama M, Isono Y, Sato T, Watanabe R, Yonekura-Sakakibara K, et al. Localization of a flavonoid biosynthetic polyphenol oxidase in vacuoles. The Plant Journal. 2006;**45**(2):133-143

[44] Dinkins RD, Hancock J, Coe BL, May JB, Goodman JP, Bass WT, et al. Isoflavone levels, nodulation and gene expression profiles of a CRISPR/Cas9 deletion mutant in the isoflavone synthase gene of red clover. Plant Cell Reports. 2021;**40**(3):517-528

[45] Zuk M, Szperlik J, Hnitecka A, Szopa J. Temporal biosynthesis of flavone constituents in flax growth stages. Plant Physiology and Biochemistry. 2019;**142**:234-245

[46] Wu J, Wang XC, Liu Y, Du H, Shu QY, Su S, et al. Flavone synthases from Lonicera japonica and L. macranthoides reveal differential flavone accumulation. Scientific Reports. 2016;**6**(1)

[47] Li H, Li D, Yang Z, Zeng Q, Luo Y, He N. Flavones produced by mulberry flavone synthase type I constitute a defense line against the ultraviolet-B stress. Plants. 2020;**9**(2):215

[48] Wang H, Liu S, Wang T, Liu H, Xu X, Chen K, et al. The moss flavone synthase I positively regulates the tolerance of plants to drought stress and UV-B radiation. Plant Science. 2020;**298**:110591

[49] Chu S, Wang J, Zhu Y, Liu S, Zhou X, Zhang H, et al. An R2R3-type MYB transcription factor, GmMYB29, regulates isoflavone biosynthesis in soybean. PLOS Genetics. 2017;**13**(5):e1006770

[50] Bosse MA, Silva MB, Oliveira NGRM, Araujo MA, Rodrigues C, Azevedo JP, et al. Physiological impact of flavonoids on nodulation and ureide metabolism in legume plants. Plant Physiology and Biochemistry. 2021;**166**:512-521

[51] Pandey A, Misra P, Khan MP, Swarnkar G, Tewari MC, Bhambhani S, et al. Co-expression ofiArabidopsis/itranscription factor,iAtMYB12/i, and soybean isoflavone synthase,iGmIFS1/i, genes in tobacco leads to enhanced biosynthesis of isoflavones and flavonols resulting in osteoprotective activity. Plant Biotechnology Journal. 2013;**12**(1):69-80

[52] Sohn SI, Pandian S, Oh YJ, Kang HJ, Cho WS, Cho YS. Metabolic engineering of Isoflavones: An updated overview. Frontiers in Plant Science. 2021:12

[53] Veremeichik GN, Grigorchuk VP, Butovets ES, Lukyanchuk LM, Brodovskaya EV, Bulgakov DV, et al. Isoflavonoid biosynthesis in cultivated and wild soybeans grown in the field under adverse climate conditions. Food Chemistry. 2021;**342**:128292

[54] Artigot MP, Daydé J, Berger M. Expression of key genes of the Isoflavonoid pathway in hypocotyls and cotyledons during soybean seed maturation. Crop Science. 2013;**53**(3):1096-1108

[55] He X, Blount JW, Ge S, Tang Y, Dixon RA. A genomic approach to isoflavone biosynthesis in kudzu (Pueraria lobata). Planta. 2011;**233**(4):843-855

[56] Vadivel AKA, McDowell T, Renaud JB, Dhaubhadel S. A combinatorial action of GmMYB176 and GmbZIP5 controls isoflavonoid biosynthesis in soybean (Glycine max). Communications Biology. 2021;**4**(1)

[57] Malla A, Shanmugaraj B, Srinivasan B, Sharma A, Ramalingam S. Metabolic engineering of Isoflavonoid Biosynthesis by expressing Glycine max Isoflavone synthase in Allium cepa L. for Genistein production. Plants. 2020;**10**(1):52

[58] Meng X, Li Y, Zhou T, Sun W, Shan X, Gao X, et al. Functional differentiation of duplicated Flavonoid 3-O-glycosyltransferases in the Flavonol and anthocyanin Biosynthesis of Freesia hybrida. Frontiers. Plant Science. 2019;**10**

[59] Chapman JM, Muday GK. Flavonols modulate lateral root emergence by scavenging reactive oxygen species in Arabidopsis thaliana. Journal of Biological Chemistry. 2021;**296**:100222

[60] Meyer P, Heidmann I, Forkmann G, Saedler H. A new petunia flower colour generated by transformation of a mutant with a maize gene. Nature. 1987;**330**(6149):677-678

[61] Shimada N, Sasaki R, Sato S, Kaneko T, Tabata S, Aoki T, et al. A comprehensive analysis of six dihydroflavonol 4-reductases encoded by a gene cluster of the Lotus japonicus genome. Journal of Experimental Botany. 2005;**56**(419):2573-2585

[62] LaFountain AM, Yuan YW. Repressors of anthocyanin biosynthesis. New Phytologist. 2021

[63] Yan H, Pei X, Zhang H, Li X, Zhang X, Zhao M, et al. MYB-mediated regulation of anthocyanin Biosynthesis. International Journal of Molecular Sciences. 2021;**22**(6):3103

[64] Xie S, Zhao T, Zhang Z, Meng J. Reduction of Dihydrokaempferol by iVitis vinfera/i Dihydroflavonol 4-reductase to produce Orange Pelargonidin-type anthocyanins. Journal

of Agricultural and Food Chemistry. 2018;**66**(13):3524-3532

[65] Solfanelli C, Poggi A, Loreti E, Alpi A, Perata P. Sucrose-specific induction of the anthocyanin biosynthetic pathway in Arabidopsis. Plant Physiology. 2005;**140**(2):637-646

[66] Naing AH, Kim CK. Roles of R2R3-MYB transcription factors in transcriptional regulation of anthocyanin biosynthesis in horticultural plants. Plant Molecular Biology. 2018;**98**(1-2):1-18

[67] Sasaki N, Nakayama T. Achievements and perspectives in biochemistry concerning anthocyanin modification for blue flower coloration. Plant and Cell Physiology. 2014;**56**(1):28-40

[68] Giampieri F, Gasparrini M, Forbes-Hernandez TY, Mazzoni L, Capocasa F, Sabbadini S, et al. Overexpression of the Anthocyanidin synthase gene in strawberry enhances antioxidant capacity and cytotoxic effects on human hepatic cancer cells. Journal of Agricultural and Food Chemistry. 2018;**66**(3):581-592

[69] Tanner GJ, Francki KT, Abrahams S, Watson JM, Larkin PJ, Ashton AR. Proanthocyanidin Biosynthesis in plants. Journal of Biological Chemistry. 2003;**278**(34):31647-31656

[70] Henry-Kirk RA, McGhie TK, Andre CM, Hellens RP, Allan AC. Transcriptional analysis of apple fruit proanthocyanidin biosynthesis. Journal of Experimental Botany. 2012;**63**(15):5437-5450

[71] Dixon RA, Xie DY, Sharma SB. Proanthocyanidins—A final frontier in flavonoid research? New Phytologist. 2004;**165**(1):9-28

[72] Grotewold E. THE GENETICS AND BIOCHEMISTRY OF FLORAL PIGMENTS. Annual Review of Plant Biology. 2006;**57**(1):761-780

[73] Yuan L, Wang L, Han Z, Jiang Y, Zhao L, Liu H, et al. Molecular cloning and characterization of PtrLAR3, a gene encoding leucoanthocyanidin reductase from Populus trichocarpa, and its constitutive expression enhances fungal resistance in transgenic plants. Journal of Experimental Botany. 2012;**63**(7):2513-2524

[74] Dong W, Ma H, Chen C, Li Y. Overexpression of the OvBAN gene enhances the proanthocyanidin content in transgenic alfalfa (Medicago sativa L.). In Vitro Cellular & Developmental Biology—Plant. 2020;**56**(4):548-557

[75] Xin Y, Meng S, Ma B, He W, He N. Mulberry genes MnANR and MnLAR confer transgenic plants with resistance to Botrytis cinerea. Plant Science. 2020;**296**:110473

[76] Dubos C, Stracke R, Grotewold E, Weisshaar B, Martin C, Lepiniec L. MYB transcription factors in Arabidopsis. Trends in Plant Science. 2010;**15**(10):573-581

[77] Agati G, Galardi C, Gravano E, Romani A, Tattini M. Flavonoid distribution in tissues of Phillyrea latifolia L. leaves as estimated by Microspectrofluorometry and multispectral fluorescence microimaging. Photochemistry and Photobiology. 2007;**76**(3):350-360

[78] Schmitz-Hoerner R, Weissenböck G. Contribution of phenolic compounds to the UV-B screening capacity of developing barley primary leaves in relation to DNA damage and repair under elevated UV-B levels. Phytochemistry. 2003;**64**(1):243-255

[79] Tattini M, Guidi L, Morassi-Bonzi L, Pinelli P, Remorini D, Degl'Innocenti E, et al. On the role of flavonoids in the integrated mechanisms of response of. New Phytologist. 2005;**167**(2):457-470

[80] Mouradov A, Spangenberg G. Flavonoids: a metabolic network mediating plants adaptation to their real estate. Frontiers. Plant Science. 2014;**5**

[81] It W-SB, Takes a Garden. How work on diverse plant species has contributed to an understanding of Flavonoid metabolism. Plant Physiology. 2001;**127**(4):1399-1404

[82] Abeynayake SW, Panter S, Chapman R, Webster T, Rochfort S, Mouradov A, et al. Biosynthesis of Proanthocyanidins in white clover flowers: Cross talk within the Flavonoid pathway. Plant Physiology. 2011;**158**(2):666-678

[83] Xie DY, Sharma SB, Paiva NL, Ferreira D, Dixon RA. Role of Anthocyanidin reductase. Encoded by. Science. 2003;**299**(5605):396-399

[84] Pang Y, Peel GJ, Wright E, Wang Z, Dixon RA. Early steps in Proanthocyanidin Biosynthesis in the model legume iMedicago truncatula/i. Plant Physiology. 2007;**145**(3):601-615

[85] Weng JK, Chapple C. The origin and evolution of lignin biosynthesis. New Phytologist. 2010;**187**(2):273-285

[86] Bvd R, Danoun S, Boudet AM, Rochange SF. Down-regulation of cinnamoyl-CoA reductase in tomato (Solanum lycopersicum L.) induces dramatic changes in soluble phenolic pools. Journal of Experimental Botany. 2006;**57**(6):1399-1411

[87] Dauwe R, Morreel K, Goeminne G, Gielen B, Rohde A, Beeumen JV, et al. Molecular phenotyping of lignin-modified tobacco reveals associated changes in cell-wall metabolism, primary metabolism, stress metabolism and photorespiration. The Plant Journal. 2007;**52**(2):263-285

[88] D'Amelia V, Aversano R, Ruggiero A, Batelli G, Appelhagen I, Dinacci C, et al. Subfunctionalization of duplicate MYB genes in iSolanum commersonii/i. Plant, Cell & Environment. 2017;**41**(5):1038-1051

[89] Bashandy H, Teeri TH. Genetically engineered orange petunias on the market. Planta. 2017;**246**(2):277-280

[90] Scarano A, Chieppa M, Santino A. Looking at Flavonoid biodiversity in horticultural crops: A colored mine with nutritional benefits. Plants. 2018;**7**(4):98

[91] Wang N, Jiang S, Zhang Z, Fang H, Xu H, Wang Y, et al. Malus sieversii: The origin, flavonoid synthesis mechanism, and breeding of red-skinned and red-fleshed apples. Horticulture. Research. 2018;**5**(1)

[92] Gao Y, Liu J, Chen Y, Tang H, Wang Y, He Y, et al. Tomato SlAN11 regulates flavonoid biosynthesis and seed dormancy by interaction with bHLH proteins but not with MYB proteins. Horticulture. Research. 2018;**5**(1)

[93] Bovy A, Schijlen E, Hall RD. Metabolic engineering of flavonoids in tomato (Solanum lycopersicum): The potential for metabolomics. Metabolomics. 2007;**3**(3)

[94] Lim W, Miller R, Park J, Park S. Consumer sensory analysis of high Flavonoid transgenic tomatoes. Journal of Food Science. 2014;**79**(6):S1212-S1217

[95] Massa S, Pagliarello R, Cemmi A, Sarcina ID, Bombarely A,

Demurtas OC, et al. Modifying anthocyanins Biosynthesis in tomato hairy roots: A test bed for plant resistance to ionizing radiation and antioxidant properties in space. Frontiers in Plant Science. 2022;**13**

[96] Di A, Manuela M, Sacco A, Frusciante L, Barone A. Genetic transformation in tomato: Novel tools to improve fruit quality and pharmaceutical production. 2011

[97] Park S, Choi M, Lee J, Kim J, Ha SH, Lim SH. Molecular and biochemical analysis of two Rice Flavonoid 3'-hydroxylase to evaluate their roles in Flavonoid Biosynthesis in Rice grain. International Journal of Molecular Sciences. 2016;**17**(9):1549

[98] Kim B, Piao R, Lee G, Koh E, Lee Y, Woo S, et al. OsCOP1 regulates embryo development and flavonoid biosynthesis in rice (Oryza sativa L.). Theoretical and Applied Genetics. 2021;**134**(8):2587-2601

[99] Ogo Y, Takaiwa F. Development of the transgenic rice accumulating flavonoids in seed by metabolic engineering. Phytonutritional Improvement of Crops. 2017:451-469

[100] Ogo Y, Mori T, Nakabayashi R, Saito K, Takaiwa F. Transgenic rice seed expressing flavonoid biosynthetic genes accumulate glycosylated and/or acylated flavonoids in protein bodies. Journal of Experimental Botany. 2015;**67**(1):95-106

[101] Cai Z, He F, Feng X, Liang T, Wang H, Ding S, et al. Transcriptomic analysis reveals important roles of lignin and Flavonoid biosynthetic pathways in Rice Thermotolerance during reproductive stage. Frontiers in Genetics. 2020;**11**

[102] Cooper M, Gho C, Leafgren R, Tang T, Messina C. Breeding drought-tolerant maize hybrids for the US corn-belt: Discovery to product. Journal of Experimental Botany. 2014;**65**(21):6191-6204

[103] Frey FP, Urbany C, Hüttel B, Reinhardt R, Stich B. Genome-wide expression profiling and phenotypic evaluation of European maize inbreds at seedling stage in response to heat stress. BMC Genomics. 2015;**16**(1)

[104] Horton DE, Johnson NC, Singh D, Swain DL, Rajaratnam B, Diffenbaugh NS. Contribution of changes in atmospheric circulation patterns to extreme temperature trends. Nature. 2015;**522**(7557):465-469

[105] Li B, Fan R, Sun G, Sun T, Fan Y, Bai S, et al. Flavonoids improve drought tolerance of maize seedlings by regulating the homeostasis of reactive oxygen species. Plant and Soil. 2021;**461**(1-2):389-405

[106] Li S, Liang T, Zhang Y, Huang K, Yang S, Lv H, et al. Vitexin alleviates high-fat diet induced brain oxidative stress and inflammation via anti-oxidant, anti-inflammatory and gut microbiota modulating properties. Free Radical Biology and Medicine. 2021;**171**:332-344

[107] Skodra C, Titeli VS, Michailidis M, Bazakos C, Ganopoulos I, Molassiotis A, et al. Olive fruit development and ripening: Break on through to the "-omics" side. International Journal of Molecular Sciences. 2021;**22**(11):5806

[108] Bian XH, Li W, Niu CF, Wei W, Hu Y, Han JQ, et al. A class B heat shock factor selected for during soybean domestication contributes to salt tolerance by promoting flavonoid biosynthesis. New Phytologist. 2019;**225**(1):268-283

[109] Zhang Y, Guo W, Chen L, Shen X, Yang H, Fang Y, et al. CRISPR/Cas9-mediated targeted mutagenesis of GmUGT enhanced soybean resistance against leaf-chewing insects through flavonoids Biosynthesis. Frontiers. Plant Science. 2022;**13**

[110] Kubra G, Khan M, Munir F, Gul A, Shah T, Hussain A, et al. Expression characterization of Flavonoid biosynthetic pathway genes and transcription factors in Peanut under water deficit conditions. Frontiers. Plant Science. 2021;**12**

[111] Ren Y, Zhang N, Li R, Ma X, Zhang L. Comparative transcriptome and flavonoids components analysis reveal the structural genes responsible for the yellow seed coat color of iBrassica rapa/i L. PeerJ. 2021;**9**:e10770

[112] Zhang L, Su W, Tao R, Zhang W, Chen J, Wu P, et al. RNA sequencing provides insights into the evolution of lettuce and the regulation of flavonoid biosynthesis. Nature. Communications. 2017;**8**(1)

[113] He R, Zhang Y, Song S, Su W, Hao Y, Liu H. UV-A and FR irradiation improves growth and nutritional properties of lettuce grown in an artificial light plant factory. Food Chemistry. 2021;**345**:128727

[114] Ismail H, Dilshad E, Waheed MT, Sajid M, Kayani WK, Mirza B. Transformation of Lactuca sativa L. with rol C gene results in increased antioxidant potential and enhanced analgesic, anti-inflammatory and antidepressant activities in vivo. 3. Biotech. 2016;**6**(2)

[115] Ismail H, Dilshad E, Waheed MT, Mirza B. Transformation of lettuce with rol ABC genes: Extracts show enhanced antioxidant, analgesic, anti-inflammatory, antidepressant, and anticoagulant activities in rats. Applied Biochemistry and Biotechnology. 2016;**181**(3):1179-1198

[116] Pink DAC, Keane EM. Lactuca: sativa L. In: Genetic Improvement of Vegetable Crops. Pergamon; 1993. pp. 543-571

[117] Matsui K, Hisano T, Yasui Y, Mori M, Walker AR, Morishita T, et al. Isolation and characterization of genes encoding leucoanthocyanidin reductase (FeLAR) and anthocyanidin reductase (FeANR) in buckwheat (Fagopyrum esculentum). Journal of Plant Physiology. 2016;**205**:41-47

[118] Zhong L, Lin Y, Wang C, Niu B, Xu Y, Zhao G, et al. Chemical profile, antimicrobial and antioxidant activity assessment of the crude extract and its Main flavonoids from Tartary buckwheat sprouts. Molecules. 2022;**27**(2):374

[119] Luthar Z, Golob A, Germ M, Vombergar B, Kreft I. Tartary buckwheat in human nutrition. Plants. 2021;**10**(4):700

[120] Spanou C, Veskoukis AS, Kerasioti T, Kontou M, Angelis A, Aligiannis N, et al. Flavonoid glycosides isolated from unique legume plant extracts as novel inhibitors of xanthine oxidase. PLoS One. 2012;**7**(3):e32214

[121] Dixon RA, Lamb CJ, Masoud S, Sewalt VJH, Paiva NL. Metabolic engineering: Prospects for crop improvement through the genetic manipulation of phenylpropanoid biosynthesis and defense responses—A review. Gene. 1996;**179**(1):61-71

[122] Liu J, Yu X, Qin Q, Dinkins RD, Zhu H. The impacts of domestication and breeding on nitrogen fixation Symbiosis

in legumes. Frontiers in Genetics. 2020;**11**

[123] Martín MA, Goya L, Pascual-Teresa S. Effect of cocoa and cocoa products on cognitive performance in young adults. Nutrients. 2020;**12**(12):3691

[124] Vauzour D, Rodriguez-Mateos A, Corona G, Oruna-Concha MJ, Spencer JPE. Polyphenols and human health: Prevention of disease and mechanisms of action. Nutrients. 2010;**2**(11):1106-1131

[125] Ofosu FK, Elahi F, Daliri EBM, Yeon SJ, Ham HJ, Kim JH, et al. Flavonoids in decorticated sorghum grains exert antioxidant, antidiabetic and Antiobesity activities. Molecules. 2020;**25**(12):2854

[126] Kim EG, Yun S, Park JR, Jang YH, Farooq M, Yun BJ, et al. Bio-efficacy of Chrysoeriol7, a natural chemical and repellent, against Brown Planthopper in Rice. International Journal of Molecular Sciences. 2022;**23**(3):1540

[127] Angeli V, Silva PM, Massuela DC, Khan MW, Hamar A, Khajehei F, et al. Quinoa (Chenopodium quinoa Willd.): An overview of the potentials of the "Golden grain" and socio-economic and environmental aspects of its cultivation and marketization. Food. 2020;**9**(2):216

[128] Zhang Y, Yan Y, Li W, Huang K, Li S, Cao H, et al. Microwaving released more polyphenols from black quinoa grains with hypoglycemic effects compared with traditional cooking methods. Journal of the Science of Food and Agriculture. 2022

[129] Lakshmanan DK, Murugesan S, Rajendran S, Ravichandran G, Elangovan A, Raju K, et al. iBrassica juncea/i (L.) Czern. Leaves alleviate adjuvant-induced rheumatoid arthritis in rats via modulating the finest disease targets—IL2RA, IL18 and VEGFA. Journal of Biomolecular Structure and Dynamics. 2021:1-14

[130] Gonzali S, Perata P. Anthocyanins from purple tomatoes as novel antioxidants to promote human health. Antioxidants. 2020;**9**(10):1017

[131] Rao MJ, Xu Y, Tang X, Huang Y, Liu J, Deng X, et al. CsCYT75B1, a citrus CYTOCHROME P450 gene, is involved in accumulation of antioxidant flavonoids and induces drought tolerance in transgenic Arabidopsis. Antioxidants. 2020;**9**(2):161

[132] Lebedev VG, Subbotina NM, MaluchenkoOP,LebedevaTN,KrutovskyKV, Shestibratov KA. Transferability and polymorphism of SSR markers located in Flavonoid pathway genes in Fragaria and Rubus species. Genes. 2019;**11**(1):11

[133] Zhai Y, Yu K, Cai S, Hu L, Amoo O, Xu L, et al. Targeted mutagenesis of. Plant Biotechnology Journal. 2020;**18**(5):1153-1168

[134] Ballester AR, Molthoff J, Vos R, BtL H, Orzaez D, Fernaੱndez-Moreno JP, et al. Biochemical and molecular analysis of Pink tomatoes: Deregulated expression of the gene encoding transcription factor SlMYB12 leads to Pink tomato fruit color. Plant Physiology. 2009;**152**(1):71-84

[135] Lin T, Zhu G, Zhang J, Xu X, Yu Q, Zheng Z, et al. Genomic analyses provide insights into the history of tomato breeding. Nature Genetics. 2014;**46**(11):1220-1226

[136] Khan S, Ullah MW, Siddique R, Nabi G, Manan S, Yousaf M, et al. Role of recombinant DNA technology to improve life. International Journal of Genomics. 2016;**2016**:1-14

[137] Colliver S, Bovy A, Collins G, Muir S, Robinson S, Vos CHR, et al. Improving the nutritional content of tomatoes through reprogramming their flavonoid biosynthetic pathway. Phytochemistry Reviews. 2002;**1**(1):113-123

[138] Lim W, Li J. Co-expression of onion chalcone isomerase in Del/Ros1-expressing tomato enhances anthocyanin and flavonol production. Plant Cell, Tissue and Organ Culture (PCTOC). 2016;**128**(1):113-124

[139] Zhou Z, Tan H, Li Q, Chen J, Gao S, Wang Y, et al. CRISPR/Cas9-mediated efficient targeted mutagenesis of RAS in salvia miltiorrhiza. Phytochemistry. 2018;**148**:63-70

[140] Li B, Cui G, Shen G, Zhan Z, Huang L, Chen J, et al. Targeted mutagenesis in the medicinal plant salvia miltiorrhiza. Scientific Reports. 2017;**7**(1)

[141] Allan AC, Espley RV. MYBs drive novel consumer traits in fruits and vegetables. Trends in Plant Science. 2018;**23**(8):693-705

[142] Balestrini R, Brunetti C, Cammareri M, Caretto S, Cavallaro V, Cominelli E, et al. Strategies to modulate specialized metabolism in Mediterranean crops: From molecular aspects to field. International Journal of Molecular Sciences. 2021;**22**(6):2887

[143] Olofsson L, Lundgren A, Brodelius PE. Trichome isolation with and without fixation using laser microdissection and pressure catapulting followed by RNA amplification: Expression of genes of terpene metabolism in apical and sub-apical trichome cells of Artemisia annua L. Plant Science. 2012;**183**:9-13

[144] Ayaz M, Sadiq A, Junaid M, Ullah F, Ovais M, Ullah I, et al. Flavonoids as prospective Neuroprotectants and their therapeutic propensity in aging associated neurological disorders. Frontiers in aging. Neuroscience. 2019;**11**

[145] Gürler SB, Kiraz Y, Baran Y. Flavonoids in cancer therapy: current and future trends. Biodiversity and Biomedicine. 2020:403-440

[146] Kay CD. The future of flavonoid research. British Journal of Nutrition. 2010;**104**(S3):S91-S95